AF213938

Studies in History and Philosophy of Science

Volume 66

Series Editor

Charles T. Wolfe, Département de philosophie & ERRAPHIS, Université de Toulouse Jean-Jaurès, Toulouse, France

Editorial Board Members

Catherine Abou-Nemeh, School of History, Philosophy, Political Science, Victoria University of Wellington, Wellington, New Zealand

Rachel A. Ankeny, University of Adelaide, Adelaide, Australia

Peter Anstey, Australian Catholic University, North Sydney, NSW, Australia

Delphine Bellis, Paul Valéry University, Montpellier, France

Meyssa Ben Saad, Université de la Manouba, Manouba, Tunisia

Hourya Bentouhami, University of Toulouse, Toulouse, France

Antonio Clericuzio, Roma Tre University, Roma, Italy

Sophia M. Connell, Birkbeck, University of London, London, UK

Matthew Daniel Eddy, Durham University, Durham, UK

Nicholas Dew, McGill University, Montreal, Canada

Ofer Gal, Unit for History and Philosophy of Science, University of Sydney, Sydney, Australia

Laura Georgescu, University of Groningen, Groningen, The Netherlands

Thierry Hoquet, Université Paris Nanterre, Nanterre, France

Clemency Montelle, School of Mathematics & Statistics, University of Canterbury, Christchurch, New Zealand

Pietro Daniel Omodeo, Ca' Foscari University of Venice, Venice, Germany

Carla Rita Palmerino, Radboud University, Nijmegen, The Netherlands

Lydia Patton, Virginia Tech, Virginia, USA

Nicholas Rasmussen, UNSW Sydney, Kensington, Australia

Jonathan Regier, Ca' Foscari University of Venice, Venice, Italy

Anne-Lise Rey, Université Paris Nanterre, Nanterre, France

Sophie Roux, République des savoirs, École Normale Supérieure - PSL, Paris, France

C. J. Schilt, Vrije Universiteit Brussel, Brussels, Belgium

Suman Seth, Cornell University, New York, USA

Tzuchien Tho, University of Bristol, Bristol, UK

Angela Willey, University of Massachusetts Amherst, Amherst, USA

Anna C. Zielinska, Département de Philosophie, Université de Lorraine & Archives Henri Poincaré, Nancy, France

Studies in History and Philosophy of Science is a peer-reviewed book series, dedicated to the history of science and historically informed philosophy of science. The series publishes original scholarship in various related areas, including new directions in epistemology and the history of knowledge within global and colonial contexts. It includes monographs, edited collections, and translations of primary sources in the English language. These cover a broad temporal spectrum, from antiquity to modernity, and all regions of the world.

Kati Kish Bar-On

Intuitionism Resocialized

Individuals, Communities, and the Evolution of Mathematical Knowledge

Springer

Kati Kish Bar-On
Science, Technology and Society Program
Massachusetts Institute of Technology
Cambridge, MA, USA

ISSN 1871-7381 ISSN 2215-1958 (electronic)
Studies in History and Philosophy of Science
ISBN 978-3-032-05838-6 ISBN 978-3-032-05839-3 (eBook)
https://doi.org/10.1007/978-3-032-05839-3

© The Editor(s) (if applicable) and The Author(s), under exclusive license to Springer Nature Switzerland AG 2025

This work is subject to copyright. All rights are solely and exclusively licensed by the Publisher, whether the whole or part of the material is concerned, specifically the rights of translation, reprinting, reuse of illustrations, recitation, broadcasting, reproduction on microfilms or in any other physical way, and transmission or information storage and retrieval, electronic adaptation, computer software, or by similar or dissimilar methodology now known or hereafter developed.
The use of general descriptive names, registered names, trademarks, service marks, etc. in this publication does not imply, even in the absence of a specific statement, that such names are exempt from the relevant protective laws and regulations and therefore free for general use.
The publisher, the authors and the editors are safe to assume that the advice and information in this book are believed to be true and accurate at the date of publication. Neither the publisher nor the authors or the editors give a warranty, expressed or implied, with respect to the material contained herein or for any errors or omissions that may have been made. The publisher remains neutral with regard to jurisdictional claims in published maps and institutional affiliations.

This Springer imprint is published by the registered company Springer Nature Switzerland AG
The registered company address is: Gewerbestrasse 11, 6330 Cham, Switzerland

If disposing of this product, please recycle the paper.

To Guy, Romi, and Yam

Preface

My first encounter with intuitionism occurred during a captivating lecture on paradoxes by Roy Cook at the fourth Summer School on Mathematical Philosophy for Female Students, hosted by the Munich Center for Mathematical Philosophy. The fact that it took an academic gathering exclusively for women in the philosophy of mathematics to shape the trajectory of my PhD (and later this book), and that I ended up focusing on a marginalized theory in the history of mathematics, is no coincidence. Make no mistake—philosophy, and the philosophy of mathematics in particular, is still a field dominated by men, and it remains wary of the notion that social elements might undermine its status as a model of precision, objectivity, and impartial standards for knowledge creation. I make no promises about changing the former, but I do aim to set the latter adrift.

Social structures, so I came to realize, shape not only how knowledge is created, shared, and evolves, but also the very rhythm of an academic's life. In this book, I explore two core questions: (1) How do interactions between individuals and their communities shape the production of knowledge? and (2) What determines why some theories flourish while others fade? These seemingly disparate questions converge in my narrative of Brouwer's intuitionism, which seeks to reintroduce the social element and the web of human relationships into the story of the most solipsistic mathematical stream.

This book was conceived at Tel Aviv University and underwent significant transformations at MIT. The intellectual environments of these institutions profoundly influenced its evolution, resulting in a work that bridges humanities, social sciences, and exact sciences. At the Cohn Institute of Tel Aviv University, I learned that history and philosophy are woven together: one cannot meaningfully engage with history without appreciating its philosophical underpinnings, nor can one tackle philosophical questions without grounding them in historical context. At MIT (and occasionally at Harvard, Northeastern, and Boston University), I gained an appreciation for the social, psychological, and political forces that shape human endeavors. This book reflects my academic journey, weaving these perspectives into a unified narrative that bears the imprint of the places and people who have shaped it.

The completion of this book owes much to the unwavering support of many brilliant minds. Menachem Fisch reignited this journey during a freezing January morning at Tatte Café in Harvard Square, urging me "not to lose sight of the book project!". Dave Kaiser backed me even before we met and stood by me through every academic twist and turn, always with a kind smile, a joke, and a thoughtful solution to every problem. Chris Featherman's ability to elevate every piece of text he touched never ceased to amaze me. He did his best to teach me how to turn a dissertation into a book. If he didn't quite succeed, it is only because of my own limitations; every mistake in this text is mine and mine alone. Leo Corry's faith in my abilities remains an enduring source of encouragement, even from thousands of miles away. Ehud Lamm, Ori Belkind, and Snait Gisis enriched this project in countless ways, even as it diverged from their own research focuses. Finally, Michael Friedman has been a true friend and an incomparable fountain of knowledge, guiding me through this turbulent academic journey and reminding me that, in the end, our greatness is ours to claim.

I am deeply grateful to Colin McClarty, Dirk Schlimm, Moritz Epple, César dos Santos, and an anonymous reviewer for their thoughtful readings of the manuscript, and to the editors and reviewers at *Science in Context*, *Philosophy of Science*, and *Synthese* for their invaluable feedback on portions of this book published during my doctoral studies. Conferences, seminars, and workshops also played a crucial role in shaping this work. For their stimulating discussions and insights, I thank Thomas Ryckman, Reviel Netz, Peter Galison, Michael Gordin, Edouard Machery, Marco Panza, Jemma Lorenat, Helen Longino, Juliet Floyd, Andy Arana, Dor Abrahamson, Michael N. Fried, and Branden Fitelson. To the audiences at Princeton, Stanford, UC Berkeley, Harvard, Pittsburgh, Chapman, Cambridge, MIT, Edinburgh, Archives Henri-Poincaré, and the Cohn Institute, thank you for your thoughtful questions, especially those I struggled to answer.

Writing a book, so it turns out, begins with mastering the art of selling your ideas. For introducing me to the nuts and bolts of book proposals, I thank Shira Shmuely and Alma Steingart. Financial support and steadfast friendships have been equally vital in this process. My gratitude goes to the Rothschild Foundation (Yad Hanadiv), the Marie Skłodowska-Curie Actions fellowship (European Commission Horizon Europe, Project number 101108426), the Science, Technology, and Society program at MIT (especially Deborah Fitzgerald, Dave Kaiser, and Paree Pinkney), and the Cohn Institute at Tel Aviv University for their generous backing. For their friendship, thought-provoking conversations, and for creating a sanctuary of support during a period when so much seemed to be unraveling, I thank Amit Goldenberg, Amir Grinstein, Michael Friedman, and Topaz Halperin.

Finally, I owe everything to my family, my true partners in this adventure. First and foremost, to Guy, who has been my rock, uprooting his life to join me on this pilgrimage halfway across the world. I have leaned on his faith in me more times than he knows, and his unwavering support has shaped this project more than he realizes. Romi and Yam, in their cuteness and adorableness, did their best to ensure this book wouldn't see the light of day. Yet, it is to them I owe the challenge of crafting something worthy of their interest and curiosity. For their love, patience, and the chaos that comes with it, I am endlessly grateful. Mutually concluding, they found the book (mostly the idea of it) to be utterly boring. You can't please everyone.

Cambridge, MA Kati Kish Bar-On

Competing Interests The author has no competing interests to declare that are relevant to the content of this manuscript.

Contents

Chapter 1
Introduction: From Dissent to Dismissal: Brouwer's Journey in Mathematical Annals

Abstract Scientific revolutions are often told as stories of triumph, yet histories of failure to revolutionize science are equally revealing. This introduction begins with L.E.J. Brouwer's attempt to re-found mathematics on intuitionistic grounds, a bold program that mobilized worldwide debates in the 1910s and 1920s but ultimately lost ground. By foregrounding this "unsuccessful" revolution, the book highlights how resistance, controversy, and eventual dismissal illuminate the social fabric of science. In dialogue with post-Kuhnian models of scientific change, Brouwer's story illustrates how intellectual projects are shaped not only by conceptual rigor but by institutional politics, group loyalties, and interpersonal tensions. The chapter frames intuitionism as more than a failed school: it is a case study in the dynamics of dissent, showing how the mathematics of the early twentieth century was forged at the crossroads of ideas and communities.

At the turn of the twentieth century, a whirlwind of transformation swept the world, leaving an indelible mark on history. Wars reshaped the world order, social movements fought for equality, and technological leaps altered the fabric of society. Aviation pioneers took flight, while groundbreaking scientific theories challenged established norms. For example, the field of psychology was transformed by the emergence of Freud's psychoanalysis and Watson's behaviorism, reshaping the understanding and treatment of human behavior. Physics was revolutionized by Albert Einstein's theory of relativity, which fundamentally altered our understanding of time, space, and gravity, moving away from the Newtonian mechanics that had dominated for centuries. Meanwhile, in biology, the rediscovery of Gregor Mendel's work on genetics at the beginning of the century ignited the field of genetics, shifting the study of heredity to a molecular level and setting the stage for modern biology. Each of these examples is a successful "breakthrough"—a discovery that solves a problem or overcomes a barrier and eventually transforms science.

However, while successful transformations often steal the spotlight, I embark on a journey in this book to illuminate a lesser-known tale—a failed revolutionary attempt in the realm of mathematics, where one mathematician, named Brouwer, aimed to solve a fundamental problem in the foundations of the discipline, sparking

© The Author(s), under exclusive license to Springer Nature Switzerland AG 2025

K. Kish Bar-On, *Intuitionism Resocialized*, Studies in History and Philosophy of Science 66, https://doi.org/10.1007/978-3-032-05839-3_1

a decade-long debate. This story, which didn't seize victory but never faded from discussion, serves as a unique lens through which we can reflect on the nuances of historical narratives and the impact of unsuccessful revolutions on our collective understanding.

Success in science is often seen by the scientific community as an advancement that resolves a challenge, shifts existing frameworks, and significantly alters our understanding. Failure, on the other hand, doesn't necessarily mean a lack of contribution. Consider Lamarck's early evolutionary theory, which incorrectly suggested that traits acquired during an organism's life could be passed on to offspring. Though ultimately replaced by Darwinian evolution, Lamarck's ideas spurred debate, influencing the development of evolutionary thought and highlighting how even unsuccessful attempts can push the boundaries of knowledge.

A relevant aspect of successes and failures in science is that they are not simply binary; they often interact in ways that drive knowledge forward. Successful theories can emerge from the ashes of failed ideas, and failed attempts can illuminate new paths for exploration. Lamarck's theory, though flawed, sparked critical debates that laid groundwork for future evolutionary theories. Thus, success and failure in science can complement and influence each other, showing that scientific progress is a continuous dialogue, one marked by ongoing discussions, where even unsuccessful efforts contribute to the broader understanding of the natural world.

Unsuccessful scientific revolutions, though often relegated to the margins of historical narratives, are vital to our comprehensive understanding of scientific evolution. They underscore the paramount importance of consensus and the rigorous process of peer review, illustrating that science progresses through a collaborative rather than solitary endeavor. These attempts frequently clash with entrenched paradigms, revealing the conservative nature of scientific communities that favor stability and stepwise innovation over radical departures. Beyond the realm of empirical data and logical constructs, the fate of these revolutions is also deeply intertwined with sociocultural currents that can either hinder or hasten scientific acceptance. The tales of these not-quite-successful ventures serve as a beacon for future scientists, embodying the courage required to question the status quo and the resilience to withstand criticism. Taken together, these elements weave a complex tapestry that portrays scientific advancement not merely as a sequence of triumphant discoveries but as a nuanced, deeply social endeavor, marked by debate, collaboration, and disagreement at times.

Some see mathematics as unique among the sciences for its potential to be revolutionized, though such revolutions are often considered less frequent compared to other fields (Ferreira and Silva 2020; Wigderson 2019). It is largely assumed that mathematics, being cumulative and axiomatic, tends to refine or build upon existing theories rather than completely overturning them (Crowe 1975). Consider calculus for example: it emerged from centuries of earlier work, including methods by Greek mathematicians like Archimedes for approximating areas and medieval studies of sequences and infinitesimals. These concepts were extended into a formal system by Newton and Leibniz, creating powerful tools for mathematics and physics. Calculus is grounded in the axioms of algebra and geometry, which provided a logical framework for its principles, such as limits, derivatives, and integrals. Rather than discarding the foundational theories of algebra and geometry, calculus built on them, offering new ways to understand motion, curves, and areas. Later refinements by mathematicians like

Cauchy and Weierstrass further formalized calculus, enhancing its rigor without disrupting its core principles, exemplifying the continuous and cumulative nature of mathematical developments. Such gradual developments are thought to provide a sense of stability, where changes are viewed more as expansions than disruptions. Unlike empirical sciences, where new experimental data can drastically alter existing paradigms, mathematics deals with abstract concepts that are generally seen as universally true and not dependent on physical evidence. As a result, mathematical truths are often perceived as enduring indefinitely, unless proven otherwise.

Progress in mathematics is often characterized by deeper insights into known theories, the discovery of connections between various domains, or the development of more comprehensive models that unify disparate areas (Thurston 1995; Weisgerber 2023). While these advances are revolutionary in scope, they may not be perceived as such because they build upon, rather than contradict, existing knowledge. Additionally, the stability of mathematical theories and the meticulous process required to verify new propositions contribute to the appearance of a slower pace of revolutionary change. The rigorous scrutiny needed for consensus and acceptance means that new ideas take time to be fully recognized and assimilated, adding to the impression that mathematics evolves more through steady progression than through abrupt upheavals.

Successful or not, every revolution attempt has historical roots. The history of mathematics is full of false starts and undertakings that came to nothing: disproved hypotheses, inadequate techniques, misguided theorizing, and much more. All of these belong in what Thomas Kuhn famously dubbed "normal science" (Kuhn 1962). Kuhn drew our attention to the constitutive role played by a science's shared normative framework in governing the day-by-day work of scientists and its scientific assessment. However, his famous book failed to account for the rationality of framework replacement. He spoke there of the pressure exerted on a framework by accumulated anomalies and their attraction to the ambitious upstarts of the field. However, in his well-known likening of paradigm shifts to gestalt switches and religious conversions, he failed to account for the interpersonal aspect of such transitions: how interactions between individuals, communities, groups, and norms affect the trajectory of a revolutionary new theory.

Kuhn's two-tier model of scientific knowledge and scientific work, and especially his focus on scientific revolutions, have given rise to a broad array of responses—some dismissive of his approach, some more constructively critical. In this book, I closely examine four post-Kuhnian accounts of scientific sea changes of the latter category, each shares at least something of Kuhn's neo-Kantianism. Two of them were designed specifically to account for framework transitions in mathematics, one to account for the history of time-space theories in physics, and one, though more general than the other three, was applied by its author to a detailed study in the history of algebra.

Yet unlike other works in the field, the idea here is to put the four accounts to the test of an *unsuccessful* twentieth century attempt to revolutionize mathematics. I am referring to the school of intuitionism, that was established by the Dutch mathematician Luitzen Egbertus Jan Brouwer, and played a major role in the debate about the foundations of mathematics from its inception.

Brouwer's intuitionism is a philosophical-mathematical approach that expands on the idea that mathematics is a human product. It characterizes mathematics as a collection of internally coherent mental constructs, offering an alternative to classical mathematics. In contrast to the Platonistic view that regards mathematical entities as independent Platonic objects, or the formalist-finitist approach that viewed mathematical entities as elements that can be reduced to formulas and axioms, intuitionism conceives them as mental constructions created in the mind. Given that these mental constructions are shaped by human creators situated in specific temporal, spatial, cultural, and societal contexts, intuitionism becomes a particularly intriguing case study for exploring the impact of group dynamics on individual mental constructions and understanding the interconnection between communal knowledge and intrasubjective knowledge.

The aim of this book is twofold. The first is to better understand by means of the four accounts why Brouwer's intuitionistic endeavor was eventually abandoned by the mathematical community. The second aim is to shine a light on specific mathematicians and how they responded to Brouwer's ideas. Through these aims, I uncover insights into how the interplay of individuals, groups, norms, and institutions shape mathematical knowledge. Each of the four models address these goals somewhat differently, and together they combine to provide a more complex understanding of such moments than any of the four alone, can offer. This book thus presents, for the first time, a full-fledged socially-oriented history of the rise and fall of Brouwer's intuitionism. In doing so, it takes an initial stride toward a comprehensive, socially oriented narrative of the history of mathematics, encapsulating the intricate connections between individual mathematicians and the communities they inhabit.

1.1 The Emergence and Demise of Brouwer's Intuitionism

First, a word of clarification is in order. This book is not about Brouwer the person. There are several insightful and comprehensive accounts of Brouwer's life, personality, and work; most notable is Dirk van Dalen's *L.E.J Brouwer—Topologist, Intuitionist, Philosopher* (van Dalen 2013). I have practically nothing to add to these compelling works. The focus of the current book is on Brouwer's intuitionistic program, and on the reactions of the community and specific individuals, to it. Having said that, I find it impossible to consider a scientist's work as completely separate from his personal and environmental circumstances. Hence, the following paragraphs will sketch, in a very brief and concise manner, the main characteristics of the rise and fall of Brouwer's intuitionism with a glimpse of Brouwer's personal life. Nevertheless, throughout the rest of the book, the personal aspects of Brouwer's life will remain mostly in the shadows.

Luitzen Egbertus Jan Brouwer was born in 1881 at Overschie, a Rotterdam neighborhood, the second-largest city in the Netherlands. After a year, the family moved to Medemblik, where Brouwer spent his early school years, and a decade later, they moved again, this time to Haarlem, the capital of the province of North-Holland. At the age of 16, Brouwer started his academic studies at the faculty of Mathematics and Sciences at the University of Amsterdam (UVA), where he first met the mathematics professor Diederik Johannes Korteweg and started to attend his lectures. The correspondence between Brouwer and his close friend Carel S. Adama van Scheltema during Brouwer's years at UVA portrays Brouwer as a sensitive and devoted young man, often prone to mood swings and occasional depression (van Dalen 2011).

After obtaining his doctorandus (a Dutch academic title equivalent to M.A. or M.Sc. degree) in 1904, Brouwer found himself torn between philosophy and mathematics as he had to reach a decision about the future direction of his scientific career. It is evident from his 1905 booklet *Life, Art and Mysticism* that Brouwer's strong philosophical views did not dissolve during his years of academic studies (Brouwer 1905). Nevertheless, he decided to pursue a mathematical dissertation under the supervision of Korteweg. The dissertation marked the turning point in Brouwer's engagement with the wider world, in the sense that Brouwer did not give up his mystic ideas, but he realized that he had to rephrase them or put them aside if he wants to make an impact as a mathematician.

Brouwer successfully defended his dissertation in February 1907. Between 1909 and 1913, Brouwer dedicated his efforts to founding modern topology, while maintaining his intuitionistic program as a side project. He introduced significant concepts as part of classical mathematics, such as that of "mapping degree" and the so-called fixed point theorem (van Atten 2020; Brouwer 1911). After being appointed full professor ordinaries in 1913 (succeeding Korteweg) and joining the *Mathematische Annalen's* editorial board in 1914, Brouwer publicly returned to his first and foremost interest and embarked on the systematic intuitionistic reconstruction of mathematics.

During the war years, Brouwer mainly reflected on the foundations of mathematics, in particular on infinite sequences, and his deliberations during that period culminated in the publication of a series of three papers under the title "Founding mathematics independently of the logical theorem of the excluded middle" (Brouwer 1918, 1919, 1923). These papers introduced the necessary tools for a systematic practice of constructive mathematics, including the intuitionistic versions of real numbers, the continuum, and elementary topology.

In 1919, an important meeting that would eventually set in motion the foundational debate took place in Engadin. Brouwer received a visit from Hermann Weyl, David Hilbert's prominent student and philosophy enthusiast who shared Brouwer's deep concern regarding the problematic foundations of mathematics. A year later, Weyl sent Brouwer a draft of his paper, titled "On the new foundational crisis in mathematics", where he refers to Brouwer as 'the revolution' (Weyl 1921, 99).

Brouwer, who greatly admired Hilbert, was extremely flattered by Weyl's support. In September 1920, Brouwer introduced his foundational program, focusing on the core problems of mathematics, before the international mathematical forum at the *Naturforscherversammlung* at Nauheim. This "meeting of natural scientists" gathered mostly German scientists to discuss physics, mathematics, and medicine for the first time after the war. Among the 2500 participants were several prominent mathematicians, such as Felix Bernstein, Emmy Noether, Fritz Noether, Hermann Weyl, George Polya, Robert Fricke, Felix Hausdorff, Kurt Hensel, and others. Brouwer gave a talk titled "Does every real number have a decimal expansion," in which he discussed the structure of the continuum. While his Nauheim lecture remained rather neutral regarding the new intuitionistic concept of choice sequence, it did contain criticism of Hilbert's theory of the solvability of mathematical problems.[1] Such a critical presentation in an international setting set the stage for the events that followed, and the publication of Weyl's advocative paper a year later made clear that Brouwer's intuitionism could no longer be ignored. Hilbert was now forced to reply, and the foundational debate had officially started (van Atten 2017; van Dalen 1995; Hesseling 2003; Mancosu 1998; McCarty 2005; Putnam and Benacerraf 1984).

Throughout the 1920s, the foundational debate received hundreds of reactions from dozens of mathematicians (such as Alfred Errera, Paul Bernays, and Abraham Fraenkel), logicians (like Walter Dubislav and Oscar Becker), and philosophers (such as Hugo Dingler and Heinrich Scholz), among others. At the same time, Brouwer continued to develop his intuitionistic program and published another series of papers entitled "On the foundations of intuitionistic mathematics" in the *Mathematische Annalen* (Brouwer 1925, 1926, 1927). The papers presented a riper and more mature view of intuitionism, including refinements and corrections of some loose ends found in Brouwer's previous trilogy. In 1924 Brouwer published an intuitionistic proof of the fundamental theorem of algebra with his doctoral student B. de Loor (Brouwer and De Loor 1924), and in 1928 he published a refined analysis of the nature of spreads (Brouwer 1928). At the same time, while reforming mathematics and performing his academic duties, Brouwer participated in a small group of philosophers, linguistics and mathematicians named the "Signific Circle" whose members were Gerrit Mannoury and Frederik van Eeden, among others (Brouwer 1946; Kirkels 2013). Between 1922 and 1926, the group regularly convened to discuss spiritual and political progress through language reform (Brouwer 1937).

Brouwer's intuitionism played a major role in the foundational debate, but Brouwer's intuitionistic ideas were poorly understood within the mathematical community. Mathematicians often confused Weyl's theory with Brouwer's theory, and they generally considered Brouwer's intuitionism as "interesting news from the border provinces" (van Dalen 2013, 335) than as critical and important developments that should occupy one's mind. Meanwhile, the debate continued to spread, reaching the rest of Europe, the US, and Russia, and in 1925 a substantial contribution to intuitionistic logic was made by the Soviet mathematician Andrey Nikolaevich Kolmogorov, where he formalized parts of Brouwer's intuitionism and presented an elaborated account of the convoluted links between intuitionism, formalism, and other schools of thought (Kolmogorov 1925).

The year 1928 is often considered the turning point of the debate and the beginning of the end for Brouwer's intuitionistic program. As mentioned, Brouwer was part of the *Mathematische Annalen's* editorial board (one of its chief editors being

[1] Hilbert maintained that every mathematical problem could be solved or rejected. As he put it: "This conviction of the solvability of every mathematical problem is a powerful incentive to the worker. We hear within us the perpetual call: There is the problem. Seek its solution. You can find it by pure reason, for in mathematics there is no ignorabimus." (Hilbert 1902, 455).

Hilbert), for 14 years now. The *Mathematische Annalen* was the most prestigious mathematics journal at the time, and being one of its editors was considered the highest recognition for a mathematician. Despite Brouwer's diligent editorial work, in 1928, Hilbert sent a request to the other chief editors, Blumenthal, Carathéodory, and Einstein, asking their permission to remove Brouwer from the journal's editorial board due to personal conflicts and mathematical differences. After extensive correspondence involving all parties, including publisher Ferdinand Springer and his legal advisors, Hilbert's request was ultimately approved, leading to Brouwer's removal from the *Annalen's* editorial board (van Dalen 1990).

The course of events had tremendously affected Brouwer, who from that point onwards refrained from publishing in the *Mathematische Annalen* and convinced his student, Arend Heyting, to do the same (Posy 1998). As can be seen in Fig. 1.1, from 1929 onwards Brouwer never regained the energy to promote intuitionism as he did before, and his subsequent publications contributed little to the further development of intuitionism (except for his short return in 1948 to introduce his work on the creating subject).[2]

Did Brouwer intend to revolutionize mathematics? Was he a man of strategic thinking, focusing on the effective ways to promote his program and devoting all his

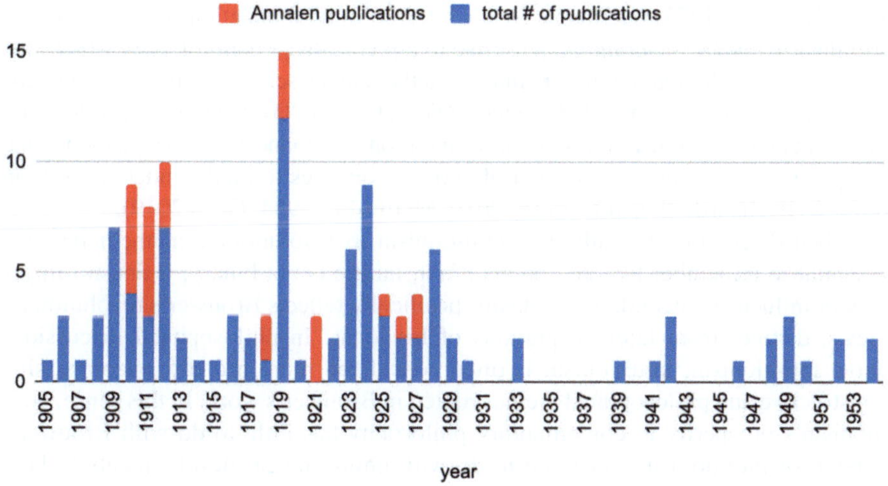

Fig. 1.1 Brouwer's publications per year from 1905 until 1955 (Annalen publications marked in red)

[2] From a broader perspective, and compared to his successors (Heyting, Kleene, Troelstra, and their successors), Brouwer produced very little intuitionistic mathematics. His work on topology was not grounded on intuition, at least not in the sense that Brouwer understood the term in the late 1910s. Between 1918 and 1928, Brouwer wrote several important papers, introducing his intuitionistic set theory and work on choice sequences, bar theorem, and fan theorem. However, from the 1930s onwards, Brouwer ceased to publish and intentionally took a step back from the foundational debate, leaving it to be continued by the next generation.

efforts to do so? The lion's share of Brouwer scholars rightfully portray Brouwer as a mathematician rather than a politician, his interest lying solely in developing his mathematical and philosophical ideas (van Atten 2004, 2020; van Dalen 2013; Hesseling 2003; van Stigt 1990). He had very few students, out of which Arend Heyting was his "only really gifted intuitionistic student,"[3] and in the 1930s, Brouwer left him to continue the intuitionistic endeavor. Heyting formalized intuitionistic logic and mathematics, broadening intuitionism's reach but diverging from Brouwer's original vision (Heyting 1930a, b, c, 1980). Although Brouwer opposed any form of axiomatization, he supported Heyting's direction and let him lead the foundational debate (van Stigt 1990). Rather than training students to carry on his own legacy, Brouwer preferred to focus on his mathematical work, allowing Heyting and others to shape the future of intuitionism.

Did Brouwer's intuitionistic program fail? It certainly did not prevail in the sense that intuitionistic notions did not replace classical ones in everyday mathematics; the antinomies of Cantor's set theory are still present, and undergraduate students encounter intuitionism only in advanced courses. On the other hand, different extensions of intuitionism continued to evolve throughout the years ranging over various disciplines.[4] Heyting's student, Anne Sjerp Troelstra, continued his supervisor's work in formalizing intuitionistic logic and choice sequences, and Michael Dummett developed a philosophical basis for intuitionism by extending Heyting's approach (Troelstra 1969; Dummett 1977). Hermann Weyl's interest in the foundations of mathematics, specifically his *The Continuum,* gave rise to a predicativist point of view which was extended by Solomon Feferman into a mathematical school of its own, namely, Predicativism (Feferman and Hellman 1995; Feferman 2005). More recent developments include semantic interpretations of intuitionism, connections to type theory and computer science, and employment of choice sequences to model indeterminacy in physics (Bezhanishvili and Holliday 2019; Martin-Löf 1984; Gisin 2019).

When discussing the failure of intuitionism to revolutionize mathematics, it's essential to remember two key points. First, failure is not binary; it doesn't imply lack of influence. Second, intuitionism specifically reflects Brouwer's revolutionary views, distinct from later adaptations of his ideas. In philosophical discussions about intuitionism, intuitionism is often considered more broadly, encompassing constructive and philosophical themes rooted in Brouwer's work. In this sense, intuitionism's prosperity in contemporary philosophy has little to do with Brouwer's version of intuitionism and more to do with philosophical developments to later versions of intuitionism.

Therefore, even though Brouwer's intuitionism did not dominate mathematics, some intuitionistic arguments never ceased to be discussed. The debate about the legitimacy of Zermelo's axiom of choice is a good example. In 1904, Ernst Zermelo introduced the axiom of choice to prove Cantor's claim that every set can be well-ordered (Zermelo 1904; Bell 2021). The introduction of the axiom provoked

[3]As documented in a letter from Brouwer to Weyl, 16.2.1928 (van Dalen 2011, 329–30).

[4]In philosophy, for example, intuitionism has a seat at the table in almost every textbook on the philosophy of mathematics. Further analysis of the philosophical community's engagement with Brouwer's intuitionism can be found in Chap. 6.

considerable criticism from the mathematical community, primarily from the French Semi-intuitionists Rene Baire, Emile Borel, and Henri Lebesgue. They argued that the axiom is not constructive as it only asserts the possibility of making several choices but does not indicate how choice functions can be defined (Diaconescu 1975). On the other hand, Hilbert regarded the axiom of choice as an essential principle of mathematics (Moore 1982). Poincaré disagreed with Hilbert, and in 1912 in a lecture on the status of set theory he delivered in Göttingen, Poincaré regarded Zermelo's line of argument as nothing but a word game that lacked all mathematical substance (Gray 2012; Rowe 2018). Brouwer commented on Zermelo's proof in his thesis in 1907, stressing that he agrees with the French semi-intuitionists' critique (Brouwer 1907, 84). The discussion around the axiom of choice eventually led to Brouwer's introduction of choice sequences (Jervell 1996).

However, the echoes of intuitionistic ideas in mainstream mathematics did not end there. Several books that were written after 1930 mention in a footnote or the preface that they are aware of the ongoing debate about the legitimacy of using the axiom of choice and made some adjustments accordingly. Van der Waerden's *Moderne Algebra* and the differences between its three editions regarding the use of the axiom of choice demonstrate the evolution of such changes. One of the major differences between the first edition that came out in 1930 (van der Waerden 1930) and the second edition in 1937 is the complete omission of the parts which rest on the axiom of choice due to van der Waerden's attempt to "avoid as much as possible any questionable set-theoretical reasoning in algebra (van der Waerden 1937, v)." However, in the third edition published in 1951 (van der Waerden 1951), the axiom of choice was reinstated. The decision to include or exclude the axiom of choice suggests that even when not accepted, intuitionistic ideas were not dismissed offhand but carefully reconsidered, at least by some mathematicians, time and again. This low-but-steady interest in Brouwer's intuitionism makes it a rather unconventional but telling case study for theories of framework transitions, especially for the four models examined in this book.

1.2 Book Contours

The structure of the book follows a deliberate progression that mirrors the historical and conceptual development of intuitionism within mathematics and philosophy. Each chapter builds on the previous one to explore how individuals and communities can drive or hinder scientific change.

Chapter 2 introduces Brouwer's intuitionism through Leo Corry's model of mathematical frameworks, analyzing the contrasting versions proposed by Brouwer, Hermann Weyl, and Arend Heyting (Corry 1989, 2001). Using Corry's distinction between the "body" and "image" of knowledge, the chapter shows how different conceptions of normative change shaped each approach. I conclude by asking whether philosophical commitments, such as Brouwer's, can motivate mathematical transformation, which leads to the next chapter's focus on the parallel connection between philosophy and science.

Chapter 3 applies Michael Friedman's model of parallel developments to Brouwer's case (Friedman 2001, 2010). Friedman argues that scientific advances are often driven by philosophical breakthroughs, using Einstein's general relativity as an example. Yet Brouwer's philosophically motivated reformation failed to gain traction. I explore this tension by comparing Brouwer's views with those of Kant, the French semi-intuitionists, and Poincaré, showing how his position diverged from each.

Chapter 4 turns to Lukas M. Verburgt's post-Kuhnian historiography, which engages both Corry's and Friedman's accounts (Verburgt 2015). Drawing on Verburgt's distinction between two kinds of *a priori*, I reinterpret Brouwer's intuitionism as a shift between images of knowledge within the same *a priori* framework. I also critically examine Verburgt's reading of intuitionism's influence on "Moscow mathematics," particularly Aleksandr Khinchin, and highlight ambiguities in his concept of epistemic breaks.

Chapter 5 shifts focus to Weyl, a key early supporter of intuitionism. Through Menachem Fisch's model of normative indecision (Fisch 2010, 2017), I explore how Weyl's philosophical ambivalence, especially concerning logical existence and the continuum, emerges from a complex process of self-deliberation and external critique. Comparing his trajectory with George Peacock's hybrid work, I analyze how different forms of conceptual hybridity affect community reception.

While the first four chapters focus on responses from the mathematical community, Chapter 6 examines how intuitionism has been received by philosophers. Through an analysis of major philosophical texts and collections, I show that intuitionism, though marginalized in mathematics, remains influential in philosophical discussions which suggests a divergent narrative of its significance.

Chapter 7 extends the analysis beyond formal structures to interpersonal and institutional dynamics. Through three case studies: (1) Weyl's relocation from Göttingen to Princeton, (2) Brouwer's engagement with the Significs group (Brouwer 1937), and (3) Brouwer's conflict with Hilbert (van Dalen 1990), I illustrate how identity, professional displacement, and interpersonal tensions shape mathematical developments. The final chapter argues that mathematical change is influenced not only by logical and conceptual innovation but also by social, psychological, and institutional factors. By tracing these dimensions, the book offers a broader framework for understanding how robust scientific knowledge emerges through the interplay of individual agency and communal norms.

References

Bell, John. 2021. The axiom of choice. In *The Stanford Encyclopedia of philosophy*. Stanford University.

Bezhanishvili, Guram, and Wesley H. Holliday. 2019. A semantic hierarchy for intuitionistic logic. *Indagationes Mathematicae* 30 (3): 403–469.

Brouwer, Luitzen Egbertus Jan. 1905. Life, art, and mysticism. *Notre Dame Journal of Formal Logic* 37:381.

Brouwer, Luitzen Egbertus Jan. 1907. *On the foundations of mathematics*. Ph.D. Dissertation, University of Amsterdam.

Brouwer, Luitzen Egbertus Jan. 1911. Über Abbildung von Mannigfaltigkeiten. *Mathematische Annalen* 71 (1): 97–115.

Brouwer, Luitzen Egbertus Jan. 1918. Founding set theory independently of the principle of the excluded middle. *KNAW Verhandelingen* 5:1–43.

Brouwer, Luitzen Egbertus Jan. 1919. Begründung Der Mengenlehre Unabhängig Vom Logischen Satz Vom Ausgeschlossenen Dritten. Zweiter Teil, Theorie Der Punktmengen. *Verhandelingen der Koninklijke Akademie van Wetenschappen te Amsterdam* 7:1–33.

Brouwer, Luitzen Egbertus Jan. 1923. Begründung Der Funktionenlehre Unabhängig Vom Logischen Satz Vom Ausgeschlossenen Dritten. Erster Teil, Stetigkeit,Messbarkeit, Derivierbarkeit. *Verhandelingen der Koninklijke Akademie van Wetenschappen te Amsterdam* 2:1–24.

Brouwer, Luitzen Egbertus Jan. 1925. Zur Begründung Der Intuitionistischen Mathematik I. *Mathematische Annalen* 93:244–257.

Brouwer, Luitzen Egbertus Jan. 1926. Zur Begründung Der Intuitionistischen Mathematik II. *Mathematische Annalen* 95:453–472.

Brouwer, Luitzen Egbertus Jan. 1927. Zur Begründung Der Intuitionistischen Mathematik III. *Mathematische Annalen* 96:451–488.

Brouwer, Luitzen Egbertus Jan. 1928. Beweis Dass Jede Menge in Einer Individualisierten Menge Enthalten Ist. *Proceedings, Koninklijke Nederlandse Akademie van Wetenschappen. Series C, Biological and Medical Sciences* 31:380–381.

Brouwer, Luitzen Egbertus Jan. 1937. Signific dialogues. In *L. E. J. Brouwer - Collected works Vol. 1 - Philosophy and foundations of mathematics*, 447–452. North-Holland Publishing.

Brouwer, Luitzen Egbertus Jan. 1946. Synopsis of the signific movement in the Netherlands. Prospects of the signific movement. In *Collected works: Philosophy and foundations of mathematics*, ed. Arend Heyting, 465–471. Elsevier Science Publishing.

Brouwer, Luitzen Egbertus Jan, and B. De Loor. 1924. Intuitionistischer Beweis Des Fundamentalsatzes Der Algebra. *Proceedings, Koninklijke Nederlandse Akademie van Wetenschappen. Series C, Biological and Medical Sciences* 27:186–188.

Corry, Leo. 1989. Linearity and reflexivity in the growth of mathematical knowledge. *Science in Context* 3 (2): 409–440.

Corry, Leo. 2001. Mathematical structures from Hilbert to Bourbaki: The evolution of an image of mathematics. In *Changing images in mathematics: From the French revolution to the new millennium*, ed. Umberto Bottazzini and Amy Dahan-Dalmédico, 167–186. Harwood Academic Publishers.

Crowe, Michael J. 1975. Ten laws concerning patterns of change in the history of mathematics. *Historia Mathematica* 2:161–166.

Diaconescu, R. 1975. Axiom of choice and complementation. *Proceedings of the American Mathematical Society* 51:176–178.

Dummett, Michael. 1977. *Elements of intuitionism*. Oxford: Clarendon Press.

Feferman, Solomon. 2005. Predicativity. In *The Oxford handbook of philosophy of mathematics and logic*, ed. Stewart Shapiro. Oxford: Oxford University Press.

Feferman, Solomon, and Geoffrey Hellman. 1995. Predicative foundations of arithmetic. *Journal of Philosophical Logic* 24:1–17.

Ferreira, Ciro, and Cibelle Silva. 2020. The roles of mathematics in the history of science: The mathematization thesis. *Transversal: International Journal for the Historiography of Science* 8.

Fisch, Menachem. 2010. Towards a history and philosophy of scientific agency. *Monist* 93 (4): 518–544.

Fisch, Menachem. 2017. *Creatively undecided: Toward a history and philosophy of scientific agency*. Chicago: The University of Chicago Press.

Friedman, Michael. 2001. *Dynamics of reason*. Stanford, CA: Center for the Study of Language and Information.

Friedman, Michael. 2010. A post-Kuhnian approach to the history and philosophy of science. *The Monist* 93 (4): 497–517.
Gisin, Nicolas. 2019. Indeterminism in physics, classical chaos and bohmian mechanics: Are real numbers really real? *Erkenntnis*. https://doi.org/10.1007/s10670-019-00165-8.
Gray, Jeremy. 2012. *Henri Poincaré: A scientific biography*. Princeton, NJ: Princeton University Press.
Hesseling, Dennis E. 2003. *Gnomes in the fog: The reception of Brouwer's intuitionism in the 1920s*. Basel, Boston, Berlin: Birkhu-ser Verlag.
Heyting, Arend. 1930a. Die Formalen Regeln Der Intuitionistischen Logik I. In *From Brouwer to Hilbert: The debate on the foundations of mathematics in the 1920s*, ed. Paolo Mancosu, 311–327. Oxford: Oxford University Press.
Heyting, Arend. 1930b. Die Formalen Regeln Der Intuitionistischen Logik II. *Sitzungsberichte der Preussischen Akademie der Wissenschaften* 1930:57–71.
Heyting, Arend. 1930c. Die Formalen Regeln Der Intuitionistischen Logik III. *Sitzungsberichte der Preussischen Akademie der Wissenschaften* 1930:158–169.
Heyting, Arend. 1980. *Collected papers*. Amsterdam: North-Holland.
Hilbert, David. 1902. Mathematical problems. *Bulletin of the American Mathematical Society* 8:437–479.
Jervell, Herman. 1996. From the axiom of choice to choice sequences. *Nordic Journal of Philosophical Logic* 1:95–98.
Kirkels, Mireille. 2013. *Mannoury's signific philosophy*. The Netherlands.
Kolmogorov, A. N. 1925. On the principle of excluded middle. In *From Frege to Godel*, ed. Jean van Heijenoort, 414–437. Cambridge, MA: Harvard University Press.
Kuhn, Thomas S. 1962. *The structure of scientific revolutions*. Chicago: University of Chicago Press.
Mancosu, Paolo, ed. 1998. *From Brouwer to Hilbert: The debate on the foundations of mathematics in the 1920s*. Oxford: Oxford University Press.
Martin-Löf, P. 1984. *Intuitionistic type theory*. Naples: Bibliopolis.
McCarty, D. C. 2005. Intuitionism in mathematics. In *The Oxford handbook of philosophy of mathematics and logic*, ed. Stewart Shapiro, 356–386. Oxford University Press.
Moore, Gregory H. 1982. Zermelo's axiom of choice: Its origins, development, and influence. *Studies in the History of Mathematics and Physical Sciences* 8:410.
Posy, Carl. 1998. Brouwer versus Hilbert: 1907-1928. *Science in Context* 11 (2): 291–325.
Putnam, Hilary, and Paul Benacerraf, eds. 1984. *Philosophy of mathematics: Selected readings*. 2nd ed. Cambridge: Cambridge University Press.
Rowe, David. 2018. *A Richer picture of mathematics: The Göttingen tradition and beyond*. Berlin: Springer.
Thurston, William P. 1995. On proof and progress in mathematics. *For the Learning of Mathematics* 15 (1): 29–37.
Troelstra, Anne Sjerp. 1969. *Principles of intuitionism*. Berlin: Springer.
van Atten, Mark. 2004. *On Brouwer*. Belmont: Wadsworth/Thomson Learning.
van Atten, Mark. 2017. The development of intuitionistic logic. In *The Stanford Encyclopedia of philosophy (Winter 2017 Edition)*. Stanford University.
van Atten, Mark. 2020. Luitzen Egbertus Jan Brouwer. In *The Stanford Encyclopedia of philosophy*. https://plato.stanford.edu/archives/win2017/entries/brouwer/. Stanford University.
van Dalen, Dirk. 1990. The war of the frogs and the mice, or the crisis of the mathematische annalen. *The Mathematical Intelligencer* 12 (4): 17–31.
van Dalen, Dirk. 1995. Hermann Weyl's intuitionistic mathematics. *The Bulletin of Symbolic Logic* 1 (2): 145–169.
van Dalen, Dirk, ed. 2011. *The selected correspondence of L.E.J. Brouwer. Sources and studies in the history of mathematics and physical science*. London: Springer.
van Dalen, Dirk. 2013. *L.E.J. Brouwer – Topologist, intuitionist, philosopher: How mathematics is rooted in life*. London: Springer.

van der Waerden, Bartel Leender. 1930. *Moderne Algebra. Unter Benutzung Der Vorlesungen von E. Artin Und E. Noether*. 1st ed. Berlin: Verlag Julius Springer.
van der Waerden, Bartel Leender. 1937. *Moderne Algebra. Unter Benutzung Der Vorlesungen von E. Artin Und E. Noether*. 2nd ed. Berlin: Verlag Julius Springer.
van der Waerden, Bartel Leender. 1951. *Moderne Algebra. Unter Benutzung Der Vorlesungen von E. Artin Und E. Noether*. 3rd ed. Berlin: Verlag Julius Springer.
van Stigt, Walter. 1990. *Studies in the history and philosophy of mathematics: Brouwer's intuitionism*. Amsterdam: North-Holland.
Verburgt, Lukas M. 2015. *"A terrible piece of bad metaphysics"? Towards a history of abstraction in nineteenth and early twentieth-century probability theory, mathematics and logic*. University of Amsterdam.
Weisgerber, Simon. 2023. Mathematical progress — On Maddy and beyond. *Philosophia Mathematica* 31 (1): 1–28. https://doi.org/10.1093/philmat/nkac019.
Weyl, Hermann. 1921. On the new foundational crisis in mathematics. In *From Brouwer to Hilbert: The debate on the foundations of mathematics in the 1920s*, ed. Paolo Mancosu, 86–118. Oxford: Oxford University Press.
Wigderson, Avi. 2019. *Mathematics and computation*. Princeton University Press. https://doi.org/10.2307/j.ctvckq7xb.
Zermelo, Ernst. 1904. Proof that every set can be well ordered. In *From Frege to Gödel: A source book in mathematical logic, 1879–1931*, ed. Jean van Heijenoort, 139–141. Cambridge, MA: Harvard University Press.

Chapter 2
Reception Unveiled: Navigating Intuitionistic Narratives

Abstract This chapter explores how Brouwer's intuitionistic program was interpreted and transformed by Hermann Weyl and Arend Heyting, each offering distinct versions of intuitionism that prompted varied responses from the mathematical community. Brouwer's vision, rooted in his philosophical ideals, sought a profound reworking of mathematics' foundations. Weyl, drawing from idealism and phenomenology, embraced Brouwer's ideas but modified them to suit his own philosophical framework, while Heyting, taking a more formal approach, restructured intuitionism in a way that made it more approachable to classical mathematicians by removing its more mystical elements. The chapter argues that the community's differing reactions to these three versions were not merely practical but reflected deeper divergences in how each mathematician viewed the normative boundaries of the discipline. Using Corry's two-tiered model of mathematical knowledge, which distinguishes between the body of knowledge and its guiding principles, the chapter shows that Heyting's formalization impacted both layers, engaging mathematicians more broadly than the more philosophically driven approaches of Brouwer and Weyl. Ultimately, the analysis suggests that radical changes to the image of mathematics are more likely to gain acceptance when they are framed in familiar terms.

Keywords Foundations of mathematics · Image of knowledge · Body of knowledge · L.E.J Brouwer · Arend Heyting · Hermann Weyl · Reflexive layer of mathematics

People react to new ideas in their own unique ways. Some are naturally intrigued about innovative and revolutionary theories, while others tend to hold on to the traditional and well-known. The story I am going to tell here revolves around a theory that was interpreted in three distinct ways by three separate individuals, leading to completely different responses from the mathematical community.[1] The range of

[1] This chapter has been previously published as: Kish Bar-On, Kati. 2023. "Connecting the Revolutionary with the Conventional: Rethinking the Differences between the Works of Brouwer, Heyting, and Weyl." *Philosophy of Science* 90 (3): 580–602. https://doi.org/10.1017/psa.2022.97

© The Author(s), under exclusive license to Springer Nature Switzerland AG 2025

K. Kish Bar-On, *Intuitionism Resocialized*, Studies in History and Philosophy of Science 66, https://doi.org/10.1007/978-3-032-05839-3_2

interpretations and varied responses reveal that the acceptance or rejection of a theory is influenced not merely by its content but also by the mindset of its audience, their commitments to the standards of the discipline, and their willingness to engage with new ideas.

Brouwer's intuitionistic program was an intriguing attempt to reform the foundations of mathematics and was probably the most controversial one within the contours of the foundational debate during the 1920s. As a whole, Brouwer's intuitionism did not prevail, but parts of the intuitionistic enterprise were developed over the years, and several mathematicians dedicated their efforts to presenting their own intuitionistic interpretations of Brouwer's work. Two of the most notable attempts are those of Hermann Weyl, whose 1921 paper brought Brouwer's intuitionism to the fore, and Arend Heyting, whose formalization of intuitionism made Brouwer's ideas more approachable to traditional mathematicians.

The mathematical community's reaction to Weyl's and Heyting's intuitionistic works was far more sympathetic than its reaction to Brouwer's original program. Most explanations for these differing reactions concern their consenting attitudes towards formal methods (Mark van Atten 2017; van Dalen 1995; Hesseling 2003) or their abandonment of Brouwer's mystical-philosophical lines of thought (Franchella 1995, 2007; Placek 1999; van Stigt 1990; Troelstra 1991). Weyl's clear and coherent writing style made his 1921 paper considerably more approachable to classical mathematicians, who often found Brouwer's papers incomprehensible and overly technical.

Weyl's compelling use of language, particularly his choice of the words 'crisis' and 'revolution', helped capture the interest of his readers.[2] Historian of science Paul Forman argued that Weyl's rhetoric was shaped by the "social-intellectual atmosphere in the aftermath of Germany's defeat" (Forman 1971, 61). He claimed that both Weyl and Hilbert "at the very least saw close parallels between the crisis in mathematics and the political crises then wracking Germany," and maintained that Germany's defeat was the main reason behind the German mathematical community's increasing interest in Brouwer's intuitionism during the 1920s (Forman 1971, 60–61). This argument is part of Forman's broader approach, known as "Forman thesis," in which he put forward the argument that the cultural values prevalent in a given place and time could affect the very content of scientific knowledge. According to Forman's thesis, the culture of Weimar Germany influenced the way many scientists, including Hermann Weyl, interpreted the concept of causality in physics and quantum mechanics (Forman 1971; Norris 2000). Forman's thesis

[2] During the 1920s, some scholars contend that it remains unclear what kind of "revolution" Hermann Weyl envisioned, while others argue that by the 1930s, Weyl's political leanings reflected those of many in the Weimar middle class who could not fathom anyone respectable aligning left of the Majority Social Democrats. Weyl himself later recalled that, like many of his contemporaries, he and even his most astute Swiss friends viewed the Communists as the primary threat to Germany in early 1930, not the rising Nazi movement. Thus, Weyl's invocation of "revolution" appears disconnected from the upheavals in Russia or even the German revolution, instead aligning him with the moderate democratic ideals championed by figures like Hilbert and the German Democratic Party.

raised an intense and controversial debate among historians and physicists that has remained at the heart of discussions about the historical relationship between science and culture ever since (Carson et al. 2011; Kraft and Kroes 1984).

The popularity of Heyting's brand of intuitionism is also often attributed to its disentanglement from Brouwer's philosophical considerations. Miriam Franchella points out that "Arend Heyting is well known not only for his own contributions to intuitionism, but also for his expository efforts to make Brouwer's ideas more accessible and more widely known [...] by separating its mystic philosophy (seen as bizarre) from its basic concepts for mathematics" (Franchella 1995, 105). In a similar vein, Thomas Placek notes that "an undeniable advantage of Heyting's approach is that it puts the fate of intuitionism on a more 'down-to-earth' ground" (Placek 1999, 106), in the sense that Heyting's intuitionism was based on considerations that were perceived as less philosophical or bizarre than Brouwer's.

Yet such explanations portray only part of the picture, as they mostly account for the practical differences between the three stories. I argue that the differences between the intuitionistic versions of Brouwer, Weyl, and Heyting run deeper. The three mathematicians held different views regarding the *normative boundaries* of the discipline that were reflected in the respective versions of intuitionism each produced. By normative boundaries I am referring to the established rules, standards, and accepted practices that define what is considered valid, credible, and appropriate within a particular field. These boundaries shape the methods, theories, and topics that are deemed acceptable for exploration and guide how practitioners within the discipline conduct research, interpret data, and present findings. They act as invisible guidelines that determine which questions are worth asking, which methodologies are suitable, and what constitutes legitimate knowledge within the field.

The following pages are devoted to a careful examination of each of the three approaches and the impact of the differences between them on the community's respective responses. I proceed by analyzing the changes suggested by Brouwer, Weyl, and Heyting in terms of changes to the practical layer of mathematics, on the one hand, and changes to the normative layer of mathematics, on the other. Such an analysis takes into account how each conceived the way mathematics should be properly done and how this manifested in their respective work.

2.1 Layering Mathematical Knowledge

In his paper "Linearity and Reflexivity in the Growth of Mathematical Knowledge," Leo Corry presents a dual-layered model of mathematics, distinguishing between a "body of knowledge" and an "image of knowledge" (Corry 1989). The body encompasses proofs, theorems, and techniques, all related to the tangible content of mathematics. The image, by contrast, consists of broader norms and guiding principles that shape what is considered valuable or worthy of pursuit, thus steering the discipline's future direction. This model draws on a framework originally proposed by

Yehuda Elkana (Elkana 1978, 1981),[3] which Corry adapts specifically to mathematics, but with a twist (Corry 2001).

The distinction between the body and image of knowledge lies in the nature of the questions they address. Questions about the adequacy, precision, and reliability of an experiment or calculation belong to the body of knowledge. Questions about what makes research beautiful, compelling, innovative, persuasive, valuable, or even risky fall under the image of knowledge. Science, therefore, evolves by grappling with two kinds of questions: those that refine the tools and techniques of discovery, and those that navigate the broader, often subjective, landscape of what is deemed important, tasteful, or groundbreaking within the field. Such a distinction is inherently fuzzy and constantly shaped by historical context, yet this doesn't imply that one layer is more important to the development of mathematics than the other.

Corry's novel insight (or twist) is that, unlike other fields, mathematics uniquely contains questions related to both the body and the image of knowledge within the discipline itself. While other sciences constantly interact between these two layers, mathematics has a unique capacity to integrate certain aspects of its image of knowledge directly into its body of knowledge. This special interconnection, which Corry refers to as "reflexive knowledge" (Corry 1989, 415), allows mathematics to turn its tools and methods inward, examining its own nature through the very framework it uses in everyday practice. Proof theory serves as a good example: it represents a doctrine that not only scrutinizes how mathematical methods should be conducted but also treats proofs as formal mathematical objects, analyzing them with mathematical techniques. Mathematics stands apart as the only discipline with such a dedicated doctrine, making its reflexive capacity truly distinctive.

With this two-tiered model in place, we can now turn to the three central figures of this chapter, namely Brouwer, Weyl, and Heyting, to examine how each reimagined the normative boundaries of mathematics, and how their respective intuitionistic programs sought to transform the body and image of mathematical knowledge.

2.2 Brouwer: Changing Normative Foundations

Brouwer's perspective on the division between the body and image of mathematical knowledge is deeply intertwined with his views on how intuitionists and formalists grapple with the concept of mathematical exactness. For Brouwer, the intuitionist

[3] Elkana proposed a compelling way to understand scientific disciplines by dividing them into two interconnected layers: the body of knowledge and the image of knowledge. The body of knowledge is the space where the actual research takes place—a realm filled with theories, concepts, and procedures that define the first-order scientific work. In contrast, the image of knowledge operates at a higher level, influencing scientific endeavors without being directly part of them. It encompasses the guiding principles that shape the discipline, such as what counts as a legitimate source of knowledge, how knowledge is validated, who the intended audience is, and how closely it aligns with the prevailing norms and ideologies of the time.

and the formalist each draw the boundaries of mathematics in distinct ways. To the intuitionist, mathematical objects are not mere symbols or formulas penned on a page; they are mental constructions, alive in the mind's creative space (Brouwer 1912, 83–84). This perception goes beyond the formalist's approach, where symbols stand detached from their meaning, existing independently as elements of a structured system. For the intuitionist, however, the meaning of mathematical relationships is inseparable from the symbols that represent them. Any alteration to these symbols transforms their meaning, rendering the idea of mathematical exactness as something that cannot rest solely on axioms and formulas.

For Brouwer, the exactness and legitimacy of a mathematical system are inextricably tied to its meaning. When meaning changes, precision shifts as well, anchoring it firmly to the image of knowledge. In Brouwer's view, questions about the nature and origins of mathematical objects clearly belong to the image of mathematical knowledge. However, he argued that not all meta-mathematical questions deserve a place within mathematics itself. Certain questions, such as why we believe in particular systems of symbolic logic or why some systems seem to capture the essence of nature better than others, reach beyond the realm of mathematics. These questions, according to Brouwer, should be explored within other disciplines, like psychology or anthropology, rather than within the image of mathematical knowledge.

Brouwer's intuitionism asserts that a mathematical object's existence hinges on its ability to be constructed within the mind. This perspective introduces a significant philosophical divide: on one side, there are objects like finite numbers and constructively defined denumerable sets, namely entities that finite beings can intuitively understand and mentally construct. On the other side lies the Cantorian collection of all real numbers, an infinite entity that surpasses our cognitive reach. Brouwer distinguished these entities by labeling the former as 'finished' or 'finishable' and the latter as 'unfinished.' A 'finished' set emerges from a recognizable, constructible process that provides a legitimate grasp of the object in its entirety, with all parts clearly determined from the initial understanding. In contrast, an 'unfinished' collection is one that we cannot fully grasp in a way that sufficiently determines all its components, leaving it perpetually incomplete in our cognitive framework (Brouwer 1952; Posy 2008).

Brouwer used this distinction to directly challenge Cantor's understanding of infinity. Cantor's theory of transfinite numbers, which differentiates between cardinality and order, introduced various classes of numbers. In this framework, ω is defined as the smallest transfinite ordinal number, representing the order type of the natural numbers under their usual linear sequence. While Brouwer accepted ω-sequences as legitimate mathematical objects (since they consist of discrete elements produced by a countably ordered process) he drew the line there. For Brouwer, mathematics could only create finite sequences based on the order type ω, and he rejected any further concept of infinite objects (Brouwer 1912, 85–86).

When it came to the set of real numbers, Brouwer was critical, referring to them as 'denumerably unfinished' (Brouwer 1907, 148–49). He argued that while, for any given denumerable subset, one could immediately find an element in the continuum that was not part of that subset, there was no positive existence claim to guarantee the existence of such an element. On this basis, Brouwer dismissed Cantor's second number class and any hierarchy of increasing cardinalities as invalid mathematical objects, challenging the foundation of Cantor's entire theory.

When it came to the intuition of the continuum, Brouwer held the firm belief that we possess an inherent understanding of the continuum as a unified whole (Brouwer 1907, 8–9). For him, the Cantorian conception of the continuum that is built from a set of independently given points was fundamentally flawed. No collection of points, no matter how exhaustive, could truly capture the nature of the continuum because, in Brouwer's view, the continuum exists as a cohesive entity unto itself, not as a collection of separable elements (Posy 2005).

Brouwer's rejection of the Cantorian continuum was mirrored in his stance on the principle of the excluded middle. He argued that this principle, which holds that any statement is either true or false, was only valid in finite systems (Brouwer 1908, 1918). In such systems, every object within the set can theoretically be examined through a finite process, allowing one to definitively determine whether a particular property, A, applies or does not apply. However, when it came to infinite systems, Brouwer maintained that this principle was unreliable. Since it is impossible to examine every object in an infinite set, the absence of a member with property A does not confirm that no such member exists. In Brouwer's view, this distinction between finite and infinite systems was critical, as it redefined the application of classical logic within mathematics.

Brouwer believed that Peano, Russell, Cantor, and Bernstein had all failed to recognize the crucial role human intuition plays when moving from finite sets to infinite ones. In finite sets, the task of examining each object could, in theory, be performed by a machine or even a trained animal. But when it came to infinite domains, human intuition became indispensable (Brouwer 1908). In Brouwer's view, mathematicians could no longer rely solely on logical rules or linguistic structures; they had to engage with the deeper, intuitive aspects of mathematical thinking.

This belief, coupled with Brouwer's insistence that mathematical objects must be constructed and his restrictive concept of infinity, led him to propose a radically new image of mathematical knowledge—one he considered the only proper way to practice mathematics. In this intuitionistic framework, the concept of mathematical truth was redefined. For Brouwer, knowing that a statement P is true required having a proof of it. To negate P meant asserting that P was refutable, that is, that a counterexample existed, but it did not imply that "not P" was provable (Brouwer 1912; Heyting 1956; Sundholm and van Atten 2008). One of the most significant implications of this reform was the rejection of proofs by contradiction, which Brouwer deemed an illegitimate method in the pursuit of mathematical knowledge.

In Brouwer's early work, leading up to 1918, he demonstrated that certain mathematical statements did not hold under intuitionistic scrutiny, and likewise, that some intuitionistic statements were not compatible with classical mathematics. However, it wasn't until he introduced the concept of choice sequences that the true incomparability between intuitionism and classical mathematics became clear (Iemhoff 2020). According to Brouwer, only potential infinity exists, meaning that infinite objects can only come into being through a process that generates them one at a time. A choice sequence represents an infinite series of numbers, elements, or finite objects created by the "free will" of the "creating subject" (Brouwer 1949; Niekus 2005). Brouwer distinguished between two types of sequences: 'lawlike' sequences, governed by a law or algorithm, and 'lawless' sequences, which operate without such constraints. An example of a lawlike sequence is the natural numbers, while a lawless sequence could be exemplified by the unpredictable results of casting a die, where, despite knowing some of the outcomes, future values remain undetermined (Dummett 1977; Fourman 1982; Kreisel 1968; Troelstra 1977).

From 1918 onward, Brouwer spoke of choice sequences as a fundamental part of his intuitionistic mathematics, which allowed him to describe the intuitionistic continuum with continuity axioms (Brouwer 1948, 1952). However, the use of the continuity axioms entailed far-reaching implications for the classical body of mathematical knowledge because classically invalid statements can derive from axioms such as the weak continuity axiom and the full axiom of continuity (Mark van Atten and Van Dalen 2002).[4]

Brouwer's intuitionistic program entailed deep and comprehensive changes to the image of knowledge, yet it did not call into question a single theory in the body of classical mathematics. No previously proven results were rejected; some were deemed unacceptable according to the new Brouwerian image of knowledge, but none were proven wrong or contradictory in consequence of Brouwer's intuitionism. Moreover, Brouwer did not use any methods from the body of knowledge to express his intuitionistic ideas. The changes Brouwer suggested affected only the image of mathematical knowledge. Nonetheless, Brouwer's ideas drew the attention of Hermann Weyl, physicist and mathematician, who was searching to find a solution to the same problem that occupied Brouwer. Before discussing Weyl's version of intuitionism, a word of clarification is in order. The versions of intuitionism suggested by Brouwer (and Weyl, as demonstrated in the next section) proposed changes to the body of mathematical knowledge that conflicted with classical reasoning, but these changes did not disprove mathematical theories. Both Brouwer and Weyl were extremely troubled by the problematic and unstable foundations of mathematics. Each suggested a somewhat different alternative to classical mathematics, which derived from the basic assumptions that mathematics is a creation of the mind and that the only legitimate mathematical objects are those obtained by construction. Hence, from their intuitionistic point of view, some classical theories were "rejected" or not valid based on being nonconstructive. However, to a non-intuitionistic mathematician, who is relatively indifferent to the paradoxes of Cantor's set theory and does not view mathematics as a mental construction, there is no (classical) logical reason to reject or invalidate any working theory. Thus the

[4] It should be noted that there are attempts to address choice sequences as supplementary to classical mathematics and not only opposing or contradicting it. For example, an approach bringing together the endlessness of time, its indeterminacy, and the free choice of the creating subject was suggested by Saul Kripke in his 2019 article, where he outlines how a concept of free-choice sequence can be combined with an acceptance of classical mathematics (Kripke 2019).

"rejected" or "unacceptable" classical theories were not rejected based on logical fallacies or mistakes but on fundamental philosophical-normative disagreements.

2.3 Hermann Weyl: Parallels and Divergences in Norms and Practice

In the early history of intuitionism, Hermann Weyl stood out as one of the few mathematicians who, at a certain point, embraced Brouwer's intuitionistic framework. Weyl's deep interest in philosophy, particularly in idealism and phenomenology, played a significant role in shaping his foundational views. The paradoxes of set theory that surfaced at the turn of the century had shaken the mathematical world, raising profound concerns about the stability of the discipline's foundations. Troubled by these paradoxes, Weyl sought an alternative to Cantor's set-theoretic framework, a journey that culminated in his book *Das Kontinuum* (Weyl 1918). This work was more than a technical response to the set-theoretical issues as it was deeply rooted in philosophical insights, particularly those drawn from phenomenology, which Weyl explicitly took as his starting point (van Dalen 2013, 309).

There has been much discussion regarding the influence of Weyl's philosophical perspectives on his foundational stance in the debates between intuitionism, constructivism, and formalism in the 1920s. Solomon Feferman argued that Weyl's critique of the set-theoretical approach led him to adopt a predicativist position in *Das Kontinuum* (Feferman 1988). Meanwhile, Erhard Scholz traced Weyl's attraction to intuitionism to philosophical considerations, particularly those inspired by Fichte's view of the continuum and space (Scholz 2000). Norman Sieroka added another layer to the conversation by suggesting that Husserl's phenomenology motivated Weyl to embrace Brouwer's intuitionism during the 1920s, while Fichte's constructivism steered him towards formalism. According to Sieroka, this explains why, after a brief alignment with Husserl's "intuitionistic-phenomenological" approach, Weyl's views shifted toward Fichte's "formalistic-constructivist" approach by 1925 (Sieroka 2009, 2019).

These philosophical shifts were reflected in Weyl's 1921 paper, *"On the New Foundational Crisis in Mathematics."* This work not only served as a clear presentation of Brouwer's intuitionistic ideas but also provided insight into Weyl's evolving approach to intuitionism (van Dalen 2013; Hesseling 2003). The paper begins by outlining the central issue, the antinomies of set theory and the "vicious circle" they created. Weyl then revisits his earlier attempt in *Das Kontinuum* to establish a firm foundation for analysis, explaining why he ultimately abandoned that project in favor of Brouwer's theory (Weyl 1921, 86). The second section delves deeper into Brouwer's ideas, covering key concepts like sequence, law, mathematical existence, real numbers, and the continuum. However, it is only towards the conclusion of the article that Weyl clearly articulates how his own intuitionistic foundations diverge from Brouwer's, marking a pivotal moment in his philosophical and mathematical journey.

Weyl acknowledged that while he didn't fully agree with Brouwer's radical conclusions, he borrowed two key ideas from him: the foundational concept of developing sequences and skepticism toward the law of the excluded middle, and the notion of *functio continua*, which refers to continuous functions (Weyl 1921, 109–110). However, Weyl expanded on these ideas by introducing his own concept of *functio mixta* and developed three central theses: (1) the concept of a sequence shifts between "law" and "choice," symbolizing the tension between "Being" and "Becoming"; (2) universal and existential theorems should be understood as abstract guidelines rather than assertions about reality; and (3) arithmetic and analysis focus on numbers and evolving sequences rather than independent theories of functions or sets.

The differences between Brouwer's and Weyl's versions of intuitionism lie primarily in the bodies of knowledge they each proposed. While both shared similar philosophical foundations, especially in their critique of classical mathematics, their approaches diverged when it came to the specific logical connections they employed and the mathematical definitions they embraced. Each had a distinct way of addressing the legitimacy of certain mathematical existence statements, which led to their respective frameworks evolving in different directions.

Despite these technical differences, their philosophical starting points in 1921 were notably aligned. Both Brouwer and Weyl shared common views on how mathematics should be practiced and were critical of the same logical principles, most notably the principle of excluded middle. This shared philosophical outlook meant that the images of mathematical knowledge they suggested, particularly regarding which general logical principles should be deemed acceptable, were strikingly similar even as their proposed methods and conclusions began to drift apart. Their journeys were marked by a tension between philosophical agreement and mathematical divergence, shaping two distinct yet philosophically linked versions of intuitionism.

Three key differences between Brouwer's and Weyl's versions of intuitionism come to light in Weyl's 1921 paper, particularly in how they approached choice sequences. First, while Brouwer allowed for both lawlike and free choice sequences, seeing each as a legitimate object, Weyl took a more restricted view (van Dalen 1995). He excluded lawlike sequences from the realm of choice sequences altogether. For Brouwer, the process by which a choice sequence was generated—whether by a fixed law or arbitrary choices—did not affect its status as a valid mathematical object (Brouwer 1948). He viewed any path leading to a sequence as equally acceptable.

Weyl, however, drew a sharp distinction between the two (Weyl 1921, 94). To him, a lawlike sequence was an object with a clear, predetermined structure, while a free choice sequence was a different kind of entity, one that was constantly evolving or "becoming." He described free choice sequences as being built step by step through arbitrary acts of choice. In his view, such a sequence could only be said to possess properties if those properties could be determined at a certain stage of its development and could not be reversed by future choices. For Weyl, a choice sequence had to be understood in its ongoing state of development, rather than as a fixed entity.

Where Brouwer treated choice sequences as established objects regardless of their construction, Weyl was deeply concerned with the process of creating them. He emphasized the importance of the choice process itself, making it central to his understanding of what a choice sequence is. In his summary of the differences between his approach and Brouwer's, Weyl clarified that when he referred to a "developing" or "becoming" choice sequence, he meant one that was lawless, free from any predetermined structure (Weyl 1921, 109–10). This distinction reflected a deeper divergence in their respective intuitionistic bodies of knowledge, with each proposing a different framework for defining and understanding the nature of choice sequences within mathematics.

The second difference between Brouwer's and Weyl's intuitionism centers on the implications of an intuitionistic approach to logic, particularly regarding existence statements. Brouwer maintained that existence statements were valid as long as they could be interpreted constructively, even if they lacked specific information. In contrast, Weyl argued that only existence statements containing concrete information were legitimate. For instance, saying "two is an even number" is valid, whereas the statement "there is an even number" lacks proper judgment because it does not specify the object in question. Despite this difference, both mathematicians agreed that constructive procedures were the only acceptable way to define existence statements—a common ground that shaped their respective images of mathematical knowledge.

A third distinction emerged in how each viewed the relationship between the image and the body of mathematical knowledge. After introducing the concept of *functio continua*, Weyl remarked that while individual functions might appear in specific theorems, they were not central to general mathematical practice. He asserted that these concepts were necessary for explaining the meaning and methods behind mathematics but were not essential to mathematics itself or its theorems. Brouwer, who had received an early draft of Weyl's paper, strongly disagreed with this point. He commented that Weyl's argument risked undermining the entire purpose of the paper. Brouwer warned that if mathematicians believed that their work was unaffected by these deeper considerations they might dismiss intuitionistic critiques altogether, and emphasized that several classical existence theorems, including those about accumulation points and minimal functions, would lose their justification under the intuitionistic framework.

This exchange highlighted a crucial difference in their thinking: Brouwer believed that mathematical tools, formulas, and methods were inseparably tied to the meaning they represented. For him, the foundational principles of intuitionism had direct consequences for all mathematical work. Weyl, on the other hand, saw a clearer distinction between the meaning of mathematics and its practice. While he acknowledged that intuitionistic definitions were important for understanding mathematical meaning, he viewed them as somewhat less relevant to the day-to-day work of mathematicians.

Despite their differences, Weyl and Brouwer were driven by similar philosophical motivations to reform the foundations of mathematics. Weyl, deeply influenced by Kant's emphasis on the primacy of intuition, saw his own philosophical journey as rooted in Kant's idea that space and time are not inherent properties of the world's objects (Weyl 1955). Brouwer shared a similar perspective, particularly on the apriority of time, dedicating a significant portion of his dissertation to refining Kant's views on the a priori (Brouwer 1907, 99–131). During the 1910s, Weyl became captivated by idealist philosophy, especially the works of Husserl and Fichte, which heavily influenced his inclination toward constructive approaches and, later, Brouwer's intuitionism (Scholz 2004; Sieroka 2009, 2019).

Because Weyl's metaphysical outlook included some realistic aspects, his idealism was different from Brouwer's idealism because Brouwer's philosophy amounted virtually to solipsism, whereas Weyl still seemed to have cleaved to phenomenology[5] (Mancosu and Ryckman 2002; Bell and Korté 2015). However, this common philosophical ground and their shared view regarding the primary intuition of time contributed to Weyl's leaning toward intuitionism. Both men shared similar views in 1921 regarding the legitimate methodologies and appropriate techniques that should be practiced within the discipline. For example, Brouwer and Weyl agreed that existential and universal statements are meaningful only if they are defined using constructive procedures (Weyl 1921, 89). In this sense, Weyl's views regarding the normative boundaries of the discipline are quite similar to Brouwer's.

Weyl's 1921 paper was an important milestone in the foundational debate. It certainly received more reactions than Brouwer's papers ever did, and it created quite a stir in the mathematical community (van Dalen 1995; Hesseling 2003). However, most reactions to Weyl's paper were written within the inner circle of Brouwer, Mannoury, Weyl, Fraenkel, and less frequently, Hilbert and Becker, between 1921 and 1926. According to the historical evidence Dennis Hesseling provides, the debate reached its peak between 1927 and 1933, and most reactions during this period were written in response to Heyting's work (Hesseling 2003, 96, 363–69). Why were mathematicians more engaged with Heyting's intuitionism than with Brouwer's or Weyl's versions of intuitionism? What is unique to Heyting's version of intuitionism?

Similar to Brouwer's intuitionism, the changes suggested by Weyl's intuitionism affected only the images of knowledge. It is true that the works of Brouwer and Weyl posed some interesting and deep questions for a small (but important) sector of the mathematical community, but they did not disprove a single theory from the classical body of knowledge, nor did they use classical methods to articulate their intuitionistic ideas. Unlike the works of Brouwer and Weyl, Heyting's formalization

[5] An extensive overview of the evolution of Weyl's idealism and the differences between Brouwer's philosophy and Weyl's phenomenological views can be found in Mancosu and Ryckman's comprehensive analysis of the correspondence between Hermann Weyl and Oscar Becker during 1923–1926 (Mancosu and Ryckman 2002). Mancosu and Ryckman show that Weyl saw a connection between intuitionism and phenomenology, and they describe how the changes in his mathematical views during the 1920s influenced his philosophical positions.

of intuitionistic mathematics entailed a different kind of change to the body and image of mathematical knowledge.

2.4 Arend Heyting: Bridging Classical Methods and Intuitionistic Mathematics

Arend Heyting was Brouwer's student at the University of Amsterdam during the 1920s. His dissertation delved deeply into the axiomatization of constructive mathematics, reflecting Brouwer's influence. In 1927, when the Dutch Mathematical Association announced a competition to formalize Brouwer's intuitionist theories, Heyting seized the opportunity. His entry not only won the prize but also paved the way for a groundbreaking essay, published in 1930 (Heyting 1930a, 1980). This work set the stage for a series of papers in which Heyting formalized intuitionistic logic and mathematics, attracting significant attention in the foundational debate of the time. His ambitious project aimed to unite intuitionistic propositional and predicate logic, arithmetic, and analysis into one cohesive system, excluding the principle of the excluded middle, a hallmark of classical logic (Heyting 1930a, b, c).

While Heyting shared Brouwer's belief that mathematical entities only exist in the human mind, he took a different stance on the role of philosophy in mathematics. For Heyting, there was a clear distinction between the practice of mathematics and its philosophical evaluation. In his view, mathematics could be constructed without philosophical preliminaries, though the value of mathematical activity might still be influenced by one's philosophical outlook (Heyting 1956, 1974). This apparent distancing from Brouwer's philosophical stance, however, should be viewed with some caution. Despite his claim that no philosophy is needed to understand mathematics, Heyting's writings reveal that intuitionistic mathematics is deeply connected to questions about meaning, introspection, and acceptable idealization (Heyting 1974, 79). In this sense, Heyting was as philosophically engaged as Brouwer, and his vision of mathematical knowledge remained thoroughly intuitionistic.

Their philosophical foundations, though, are distinct. Much of Brouwer's philosophical perspective, particularly his concept of the a priori, had roots in Kant's philosophy (van Dalen 1978; van Stigt 1990). Brouwer sought to update and refine Kant's ideas on apriority in his dissertation, aiming to bring them in line with contemporary thought (Brouwer 1907, 99). Heyting's intuitionistic mathematics was much less influenced by Kant, and many of Brouwer's core concepts, such as consciousness and the mind, did not factor into Heyting's approach (Heyting 1974). While they might have disagreed on what constituted intuitive clarity in mathematics—Heyting once remarked, "it has proved not to be so intuitively clear what is intuitively clear in mathematics" (Heyting 1962, 195)—they both stood united in their belief that mathematical entities exist only within the mind.

Unlike his mentor, Heyting adopted a more formalist approach in practice. While Brouwer resisted formal methods, Heyting viewed them as legitimate tools for expressing intuitionistic ideas within the broader framework of mathematical knowledge (van Stigt 1990, 290). In one of his final papers on intuitionism, written in 1974, Heyting outlined the dual objectives of formalizing intuitionism (Heyting 1974, 89–90). First, he saw it as an effective method for examining the assumptions underlying a given proof, noting its successful application to proofs in Brouwer's theory of choice sequences. Second, formalization allowed for the expression of intuitionistic theorems in a language that traditional mathematicians could understand.

For Heyting, formalization was a practical tool that served the purpose of advancing intuitionistic theories.[6] His approach to axiomatic methods was functional—a means to clarify and introduce Brouwer's intuitionistic ideas. By trimming down classical axiomatic systems to align with the intuitionistic viewpoint, Heyting managed to create a bridge between classical mathematicians and intuitionistic mathematics (Mark van Atten 2017). He acknowledged that while formal methods were not strictly necessary, they were highly effective in conveying the ideas and proofs of intuitionism to a broader audience (Heyting 1959, 70).

In Heyting's works, we notice a different kind of change to the body and image of knowledge than that applied in the intuitionistic versions of Brouwer and Weyl. The change Heyting suggested starts from the image of knowledge because he viewed mathematics as a mental activity. However, unlike Brouwer and Weyl, Heyting used the classical body of mathematical knowledge to express his intuitionistic ideas: he presented the philosophical views underlying intuitionistic mathematics through formal axioms. Heyting created a new theory using methods and tools from within the body of knowledge; namely, he used mathematics' unique ability to absorb a normative view of how mathematics should be properly done into its own body of knowledge.

---

[6] Although Heyting does not provide any direct references or examples of his remark regarding the successful applications of formalized methods to intuitionism in his 1974 paper, he is most likely referring to the works of Stephen Kleene, Richard Vesley, John Myhill, Georg Kreisel, and Anne Sjerp Troelstra. In 1950 Kleene started to work on an axiomatic system of intuitionistic analysis based on a language with variables for numbers and choice sequences and included arithmetical axioms, an axiom of countable choice, the axiom of bar induction, and a continuity axiom (Kleene and Vesley 1965). In the 1960s, Kreisel developed an axiomatization for lawlike sequences and choice sequences, and Myhill presented an axiom intended to express Brouwer's theory of the creative subject (Kreisel and Troelstra 1970; Myhill 1966). In 1968 Kreisel presented an axiomatization of the theory of lawless sequences, later used by Troelstra as a tool for studying other notions of choice sequence (Kreisel 1968; Troelstra 1983). According to Troelstra, the advancement in formalizing intuitionism that took place during 1963–1980 made clear that "many different notions of choice sequence may be distinguished, with different properties" (Troelstra 1991, 20).

2.5 Reflexivity, Intuitionism, and Body-Images

To better understand the differences between the intuitionistic approaches of Brouwer, Weyl, and Heyting, it's helpful to revisit Leo Corry's two-tiered model. As previously noted, mathematics is unique in its relationship between the image and body of knowledge. Corry introduces a third, intermediate layer, which bridges these two realms, known as "reflexive thinking" (Corry 1989, 425). This layer belongs to the body of mathematical knowledge, as it is produced and justified by proof like any other mathematical result. However, it also addresses second-order problems, making it connected to the images of knowledge. In Corry's view, mathematical knowledge encompasses all these layers, which engage in an ongoing dialectical debate, a process essential to understanding the historical development of mathematics.

Through this lens, Heyting's formalization of intuitionism can be seen as part of this reflexive layer. His use of formal methods to express a philosophical view of mathematical entities as mental constructs created a work that sits within both the body and image of knowledge. On the one hand, it contributed to mathematical theory; on the other, it addressed a normative question about how mathematics should be practiced.

In clear contrast, Brouwer's and Weyl's versions of intuitionism cannot be considered part of this reflexive layer of thought. The changes introduced by both Brouwer and Weyl are changes that concentrate primarily on "second-order problems," namely, on the layer of the images of knowledge. Unlike Heyting's intuitionistic mathematics, which affected and involved the body of knowledge as well, the intuitionistic versions suggested by Brouwer and Weyl were confined solely to the normative realm of the images of knowledge. Neither Brouwer nor Weyl used any methods or theories from the existing body of knowledge to express their intuitionistic ideas. Their point of departure was a normative conception of how mathematics should be properly done, but they did not use formal methods to articulate their intuitionistic approaches. According to Walter van Stigt, even though Brouwer encouraged Heyting to engage with the formalization of intuitionism, Brouwer himself opposed formalization, dismissing it as a "sterile and unproductive mathematical exercise" (van Stigt 1990, 285).

Heyting's use of formal methods as tools of communication to convey his intuitionistic ideas serves as a compelling example of the unique connection between mathematics' body and image of knowledge. Heyting's intuitionism was rooted in normative considerations, but he also believed that casting them in formal garb rendered some intuitionistic notions clearer and more accessible to the working mathematician (Heyting 1930a, 311).

Heyting's formalization put intuitionistic logic and classical logic side by side, as two parallel systems, a move that encouraged mathematicians and logicians to develop and analyze the intuitionistic system as well and compare between the two. Indeed, mathematicians were much more inclined to engage with Heyting's work. Mathematician Alfred Errera and logician Marcel Barzin embarked on a public discussion with Heyting through a series of papers in the *L'Enseignement mathematique* journal, addressing different aspects of Heyting's formalization papers (Barzin and Errera 1931, 1932a, b, 1933). Their correspondence shone a spotlight on Heyting's work and, in turn, encouraged more mathematicians to interact with it. A quantitative inquiry of the public reactions to intuitionism presented by Dennis Hesseling indicates that in the 1930s, there was a shift in the trajectory and content of the debates; the number of reactions to intuitionism in the 1930s was substantially higher than in the 1920s or earlier. These reactions mainly engaged Heyting's formalization of logic and not Brouwer's original intuitionistic program (Hesseling 2003, 93–96, 363–69).

A substantial amount of the reactions to Heyting's formalization go beyond the intuitionistic system itself and discuss the differences between classical and intuitionistic mathematics. Mathematician Kurt Gödel and logician and mathematician Gerhard Gentzen each arrived independently at a theorem proving that Peano arithmetic is translatable into Heyting arithmetic (Gödel 1933; Gentzen withdrew his paper when he learned of Gödel's paper). Gödel's paper concludes that the system of intuitionistic arithmetic and number theory is only apparently narrower than the classical one; it contains it, but with a somewhat different interpretation (Gödel 1933, 295). Thus, it seems that Gödel saw the differences between intuitionism and classical mathematics mostly as differences in the image, not the body, of knowledge.

Heyting replied to Gödel's paper, stating that for the intuitionist, it is the interpretation that is essential (Heyting 1934, 18), and his response made Gödel "supersensitive about differences in meaning" (Kreisel 1987, 82; Mark van Atten 2017). That is not to imply that Gödel was converted to intuitionism; he certainly was not. But the developments made by Gödel, Gentzen, and Karl Menger to Heyting's formalized system suggest that Heyting's formalization of intuitionism allowed for ideas belonging to the intuitionistic image of knowledge to be discussed from within the classical body of knowledge, a discussion that could not have taken place within the contours of Brouwer's or Weyl's intuitionistic versions. In order to engage with Brouwer's intuitionism or Weyl's intuitionism, mathematicians had to be acquainted, at least to some extent, with their philosophical considerations and understand the implications of such considerations for mathematics. Heyting's intuitionism offered mathematicians a way to explore intuitionistic ideas without an initial immersion in philosophical commitments they often found foreign and unrelatable.

2.6 Concluding Remarks

By revisiting Corry's two-tiered model in the context of Brouwer, Weyl, and Heyting's works, this chapter refines the lens through which we interpret the divergences in their approaches and how the mathematical community reacted to Heyting's formalization of intuitionism. Some argue that Heyting's more accessible, non-philosophical approach appealed to practicing mathematicians by avoiding the distractions of speculative philosophy (Placek 1999, 106). Others believe that it was Heyting's version of intuitionism that diverged fundamentally from Brouwer's orthodoxy which made it more popular (van Stigt 1990, 275). Yet, this chapter suggests there is more to Heyting's intuitionism than mere formalization. The philosophical motivation behind his approach, though different from Brouwer's, was rooted in a normative vision of how mathematics should be practiced. In formalizing intuitionistic ideas, Heyting was not abandoning philosophy but using formal methods as a vehicle to express a vision of mathematics that remained faithful to his ideals. As Heyting himself later admitted, the formalized system that garnered so much attention in the 1930s had, in some ways, overshadowed the very ideas it was meant to communicate (Heyting 1978, 15).

This chapter remains loyal to Heyting's intentions, positioning his use of formal methods as both a philosophical and practical strategy. His intuitionism succeeded where Brouwer's and Weyl's struggled to gain traction because it bridged the classical and intuitionistic worlds. Heyting's approach allowed the classical body of mathematics to engage with intuitionistic ideas without requiring an immersion in philosophical debate, creating a unique intersection where both layers interacted. Among Brouwer, Weyl, and Heyting, only Heyting used formal methods from the classical mathematical tradition, making his intuitionism more accessible to a wider audience and allowing for dialogue between classical and intuitionistic perspectives.

Reflecting on these stories through Corry's model raises important questions about the interconnectedness of the body and image of mathematics. Corry contends that while they are interlinked, changes in one do not necessarily cause changes in the other (Corry 2001). Lukas Verburgt, however, challenges this notion, arguing that transitions between images of knowledge often lead to transformative shifts in the body of knowledge itself, as seen in his study of modern probability theory (Verburgt 2015). In the cases of Brouwer, Weyl, and Heyting, their vision of mathematics as a mental construct inherently led to changes in both the image and body of mathematical knowledge, excluding non-constructed concepts from the mathematical canon.

While Verburgt's example focuses on a successful community-wide transition, Brouwer, Weyl, and Heyting's attempts were less successful. These stories—when placed alongside Verburgt's challenge to Corry—pose two vital questions: What is the relationship between shifts in the image of knowledge and the body of knowledge? Was Brouwer's program perhaps too philosophical to resonate with the mathematical community? The next chapter will explore these questions through Michael Friedman's model of scientific framework transitions, shedding light on the complex relationship between philosophical development and changes in mathematical practice.

References

Barzin, Marcel, and Alfred Errera. 1931. Sur La Logique de M. Heyting. *L'Enseignement Mathematique* 30:248–250.

Barzin, Marcel, and Alfred Errera. 1932a. Note Sur La Logique de M. Heyting. *L'Enseignement Mathematique* 31:122–124.

Barzin, Marcel, and Alfred Errera. 1932b. Reponse a M. Heyting. *L'Enseignement Mathematique* 31:273–274.

Barzin, Marcel, and Alfred Errera. 1933. La Logique de M. Brouwer, Etat de La Question. *Bulletin Mathematique de la Societe Roumaine des Sciences* 35:51–52.

Bell, John, and Herbert Korté. 2015. Hermann Weyl. In *The Stanford Encyclopedia of Philosophy*. Stanford University.

Brouwer, Luitzen Egbertus Jan. 1907. *On the foundations of mathematics*. Ph.D. Dissertation, University of Amsterdam.

Brouwer, Luitzen Egbertus Jan. 1908. The unreliability of logical principles. In *L.E.J Brouwer collected works I—Philosophy and foundations of mathematics*, ed. Arend Heyting, 105–111. Amsterdam: North-Holland.

Brouwer, Luitzen Egbertus Jan. 1912. Intuitionism and formalism. In *L.E.J Brouwer collected works I—Philosophy and foundations of mathematics*, ed. Arend Heyting, 123–138. Amsterdam: North-Holland.

Brouwer, Luitzen Egbertus Jan. 1918. Founding set theory independently of the principle of the excluded middle. *KNAW Verhandelingen* 5:1–43.

Brouwer, Luitzen Egbertus Jan. 1948. Essentially negative properties. *Indagationes Mathematicae* 10:322–323.

Brouwer, Luitzen Egbertus Jan. 1949. *The non-equivalence of the constructive and the negative order relation in the continuum.*

Brouwer, Luitzen Egbertus Jan. 1952. Historical background, principles and methods of intuitionism. In *L.E.J Brouwer Collected Works I—Philosophy and Foundations of Mathematics*, ed. Arend Heyting, 508–515. Amsterdam: North-Holland.

Carson, Cathryn, Alexei Kojevnikov, and Helmuth Trischler. 2011. The Forman thesis: 40 years after. In *Weimar culture and quantum mechanics*, 1–6. Co-published with Imperial College Press. https://doi.org/10.1142/9789814293129_0001.

Corry, Leo. 1989. Linearity and reflexivity in the growth of mathematical knowledge. *Science in Context* 3 (2): 409–440.

Corry, Leo. 2001. Mathematical structures from Hilbert to Bourbaki: The evolution of an image of mathematics. In *Changing images in mathematics: From the French revolution to the new millennium*, ed. Umberto Bottazzini and Amy Dahan-Dalmédico, 167–186. Harwood Academic Publishers.

Dummett, Michael. 1977. *Elements of intuitionism*. Oxford, England: Clarendon Press.

Elkana, Yehuda. 1978. *Two-tier thinking: Philosophical realism and historical relativism*. Vol. 8, 309–326. London and Beverly Hills: Sage.

Elkana, Yehuda. 1981. A programmatic attempt at an anthropology of knowledge. In *Sciences and cultures*, Sociology of the sciences yearbook, ed. Everett Mendelson and Yehuda Elkana, 1–76. Springer.

Feferman, Solomon. 1988. The development of programs for the foundations of mathematics in the first third of the 20th century. https://pdfs.semanticscholar.org/5a3a/112929a5a2a1f1fec4447cc584628918c0fb.pdf.

Forman, Paul. 1971. Weimar culture, causality, and quantum theory, 1918-1927: Adaptation by German physicists and mathematicians to a hostile intellectual environment. *Historical Studies in the Physical Sciences* 3:1–115.

Fourman, Michael P. 1982. Notions of choice sequence. In *Studies in logic and the foundations of mathematics*, vol. 110, 91–105. Elsevier.

Franchella, Miriam. 1995. Like a bee on a windowpane: Heyting's reflections on solipsism. *Synthese* 105:207–251.
Franchella, Miriam. 2007. Philosophies of intuitionism: Why we need them. *Teorema: Revista Internacional de Filosofia* 26:73–82.
Gödel, Kurt. 1933. On intuitionistic arithmetic and number theory. In *Collected works: Volume I: Publications 1929-1936*, ed. Solomon Feferman, John Dawson, Stephen Cole Kleene, Gregory H. Moore, Robert Solovay, and Jean van Heijenoort, 286–295. Oxford: Oxford University Press.
Hesseling, Dennis E. 2003. *Gnomes in the fog: The reception of Brouwer's intuitionism in the 1920s*. Basel, Boston, Berlin: Birkhäuser Verlag.
Heyting, Arend. 1930a. Die Formalen Regeln Der Intuitionistischen Logik I. In *From Brouwer to Hilbert: The debate on the foundations of mathematics in the 1920s*, ed. Paolo Mancosu, 311–327. Oxford: Oxford University Press.
Heyting, Arend. 1930b. Die Formalen Regeln Der Intuitionistischen Logik II. *Sitzungsberichte der Preussischen Akademie der Wissenschaften* 1930:57–71.
Heyting, Arend. 1930c. Die Formalen Regeln Der Intuitionistischen Logik III. *Sitzungsberichte der Preussischen Akademie der Wissenschaften* 1930:158–169.
Heyting, Arend. 1934. *Mathematische Grundlagenforschung Intuitionismus Beweistheorie*. Berlin, Heidelberg: Springer.
Heyting, Arend. 1956. *Intuitionism: An introduction*. Amsterdam: North-Holland.
Heyting, Arend. 1959. Some remarks on intuitionism. In *Constructivity in mathematics*, ed. Arend Heyting, 67–71. Amsterdam: North-Holland.
Heyting, Arend. 1962. After thirty years. In *Logic, methodology and philosophy of science*, ed. E. Nagel, P. Suppes, and A. Tarski, 194–197. Stanford: Stanford University Press.
Heyting, Arend. 1974. Intuitionistic views on the nature of mathematics. *Synthese* 27 (1–2): 79–91.
Heyting, Arend. 1978. History of the foundations of mathematics. *Nieuw Archief voor Wiskunde* 26 (3): 1–21.
Heyting, Arend. 1980. *Collected papers*. Amsterdam: North-Holland.
Iemhoff, Rosalie. 2020. Intuitionism in the philosophy of mathematics. In *The Stanford Encyclopedia of Philosophy (Winter 2019 Edition)*. Stanford University.
Kleene, Stephen Cole, and Richard Vesley. 1965. *The foundations of intuitionistic mathematics*. Amsterdam: North-Holland.
Kraft, P., and P. Kroes. 1984. Adaptation of scientific knowledge to an intellectual environment. Paul Forman's 'Weimar culture, causality, and quantum theory, 1918-1927': Analysis and review. *Centaurus* 27:76–99.
Kreisel, Georg. 1968. Lawless Sequences of Natural Numbers. *Compositio Mathematica* 20:222–248.
Kreisel, Georg. 1987. Gödel's excursions into intuitionistic logic. In *Gödel remembered*, ed. P. Weingartner and L. Schmetterer, 67–179. Napoli: Bibliopolis.
Kreisel, Georg, and Anne Sjerp Troelstra. 1970. Formal systems for some branches of intuitionistic analysis. *Annals of Mathematical Logic* 1:229–387.
Kripke, Saul. 2019. Free choice sequences: A temporal interpretation compatible with acceptance of classical mathematics. *Indagationes Mathematicae* 30 (3): 492–499.
Mancosu, Paolo, and Thomas Ryckman. 2002. Mathematics and phenomenology: The correspondence between O. Becker and H. Weyl. *Philosophia Mathematica* 10 (2): 130–202.
Myhill, John. 1966. Notes towards an axiomatization of intuitionistic analysis. *Logique et Analyse* 9 (35–36): 280–297.
Niekus, Joop. 2005. Individual choice sequences in the work of LEJ Brouwer. *Philosophia Scientiæ* 9 (S2): 217–232.
Norris, Christopher. 2000. *Quantum theory and the flight from realism: Philosophical responses to quantum mechanics*. London: Routledge.
Placek, Tomasz. 1999. *Mathematical intuitionism and Intersubjectivity: A critical exposition of arguments for intuitionism*. Dordrecht: Springer.

Posy, Carl. 2005. Intuitionism and philosophy. In *The Oxford handbook of philosophy of mathematics and logic*, ed. Stewart Shapiro, 318–355. Oxford University Press.

Posy, Carl. 2008. Brouwerian infinity. In *One hundred years of intuitionism (1907–2007)*, ed. M. van Atten, P. Boldini, M. Bourdeau, and G. Heinzmann, 21–36. Birkhäuser Basel: Publications of the Henri Poincaré Archives.

Scholz, Erhard. 2000. Herman Weyl on the concept of continuum. In *Proof theory: History and philosophical significance*, ed. V. F. Hendricks et al., 195–220. Dordrecht: Kluwer.

Scholz, Erhard. 2004. Philosophy as a cultural resource and medium of reflection for Hermann Weyl. https://arxiv.org/abs/math/0409596

Sieroka, Norman. 2009. Husserlian and Fichtean leanings: Weyl on Logicism, intuitionism, and formalism. *Philosophia Scientiae* 13 (2): 85–96.

Sieroka, Norman. 2019. Neighbourhoods and Intersubjectivity: Analogies between Weyl's analyses of the continuum and transcendental-phenomenological theories of subjectivity. In *Weyl and the problem of space: From science to philosophy*, ed. Julien Bernard and Carlos Lobo, 98–122. Switzerland: Springer.

Sundholm, Goran, and Mark van Atten. 2008. The proper explanation of intuitionistic logic: On Brouwer's demonstration of the Bar theorem. In *One hundred years of intuitionism (1907–2007)*, ed. M. van Atten, P. Boldini, M. Bourdeau, and G. Heinzmann, 60–77. Birkhäuser Basel.

Troelstra, Anne Sjerp. 1977. *Choice sequences*. Clarendon Press.

Troelstra, Anne Sjerp. 1983. Analyzing Choice Sequences. *The Journal of Philosophical Logic* 12:197–260.

Troelstra, Anne Sjerp. 1991. *A history of constructivism in the 20th century*. University of Amsterdam, ITLI Prepublication Series ML-91-05.

van Atten, Mark. 2017. The development of intuitionistic logic. In *The Stanford Encyclopedia of Philosophy (Winter 2017 Edition)*. Stanford University.

van Atten, Mark, and Dirk Van Dalen. 2002. Arguments for the continuity principle. *Bulletin of Symbolic Logic* 8 (3): 329–347.

van Dalen, Dirk. 1978. Brouwer: The genesis of his intuitionism. *Dialectica* 32:291–303.

van Dalen, Dirk. 1995. Hermann Weyl's intuitionistic mathematics. *The Bulletin of Symbolic Logic* 1 (2): 145–169.

van Dalen, Dirk. 2013. *L.E.J. Brouwer—Topologist, intuitionist, philosopher: How mathematics is rooted in life*. London: Springer.

van Stigt, Walter. 1990. *Studies in the history and philosophy of mathematics: Brouwer's intuitionism*. Amsterdam: North-Holland.

Verburgt, Lukas M. 2015. *"A terrible piece of bad metaphysics"? Towards a history of abstraction in nineteenth and early twentieth-century probability theory, mathematics and logic*. University of Amsterdam.

Weyl, Hermann. 1918. *The continuum: A critical examination of the foundation of analysis*. Thomas Jefferson University Press.

Weyl, Hermann. 1921. On the new foundational crisis in mathematics. In *From Brouwer to Hilbert: The debate on the foundations of mathematics in the 1920s*, ed. Paolo Mancosu, 86–118. Oxford: Oxford University Press.

Weyl, Hermann. 1955. Insight and reflection. In *The Spirit and uses of the mathematical sciences*, ed. T. L. Saaty and F. J. Weyl, 281–301. New York: McGraw-Hill.

Chapter 3
Ideas Unleashed: Philosophical Catalysts for Mathematical Revolution

Abstract This chapter explores the complex relationship between philosophical thought and scientific revolutions, with a focus on Michael Friedman's theory of scientific philosophy. Friedman's theory is particularly useful for understanding successful revolutions, such as Einstein's theory of relativity. However, the chapter turns to Brouwer's intuitionistic program as an intriguing counterexample—a radical mathematical and philosophical reform that, despite its intellectual promise, did not succeed in gaining widespread acceptance. Brouwer's intuitionism, inspired by the philosophical ideas of Kant, the French semi-intuitionists, and Poincaré, sought to address deep foundational issues in mathematics. The chapter analyzes how these influences shaped Brouwer's thinking and why his ideas initially attracted figures like Abraham Fraenkel and Hermann Weyl, both of whom engaged seriously with intuitionism and helped disseminate Brouwer's concepts. Yet, despite its philosophical depth and potential to solve critical problems, Brouwer's intuitionism was eventually set aside by the mathematical community. This chapter's key takeaway is that, according to Friedman's model, one reason for this rejection lies in the absence of strong philosophical underpinnings that could have provided a stable foundation for Brouwer's ideas. While Brouwer's intuitionism posed viable solutions to significant issues, its philosophical isolation left it vulnerable. The chapter concludes by highlighting the limits of Friedman's theory in explaining why some revolutionary ideas, despite their appeal, fail to gain traction.

Keywords Intuitionism · Brouwer's philosophy of mathematics · Framework transitions · Foundations of mathematics · Scientific philosophy

At the end of the last chapter, we were left with some pressing philosophical questions surrounding Brouwer's intuitionistic theory. The issues raised there—about the essential relationship between philosophical thought and scientific progress—open the door to a larger inquiry, one that this chapter takes as its focus.[1] Is there

[1] This chapter has been previously published as: Kish Bar-On, Kati. 2022. "From Philosophical Traditions to Scientific Developments: Reconsidering the Response to Brouwer's Intuitionism."

© The Author(s), under exclusive license to Springer Nature Switzerland AG 2025

K. Kish Bar-On, *Intuitionism Resocialized*, Studies in History and Philosophy of Science 66, https://doi.org/10.1007/978-3-032-05839-3_3

always a necessary connection between philosophical advancements and the evolution of science? And, if so, how should we understand that connection when reflecting on past efforts to revolutionize a field? More specifically, what kind of relationship exists between philosophy and mathematics, and why is this connection so crucial for understanding the shifts in scientific paradigms?

In exploring these questions, we turn to Michael Friedman's influential theory of scientific philosophy. In a series of lectures he gave at Stanford in 1999, which later became his book *Dynamics of Reason (*Friedman 2001*)*, Friedman lays out a compelling argument about how philosophy plays a central role in making scientific revolutions rational. Friedman suggests that philosophy provides the groundwork for radical shifts in scientific thought by generating the new ideas necessary to build fresh frameworks, ones that eventually replace the old. But can this same reasoning apply to every attempt to radically alter a scientific discipline?

Friedman's examples are fascinating, but they focus on successful revolutions. What about unsuccessful revolutions? This is where Brouwer's intuitionistic program presents an intriguing counterpoint. As we saw in the previous chapter, Brouwer's intuitionism was as much a philosophical reform of mathematics as it was a technical one. Yet despite the ambition behind it, intuitionism didn't take hold in the way that Einstein's theory of relativity did. So, the question remains: why didn't Brouwer's program succeed, particularly in light of Friedman's model?

To better understand Brouwer's motivations, we need to consider the philosophical influences that shaped his thinking. Brouwer was well-versed in the ideas of Kant and Poincaré, both of whom held significant views on mathematics and its nature. This chapter will explore how these philosophical developments may have inspired Brouwer's intuitionism, asking to what degree his work can be seen as an extension of these earlier traditions, or if it diverges too much to make a direct connection.

Brouwer's intuitionism left its mark on the history of mathematics. Throughout the 1920s, it played a significant role in the heated debates over the foundations of mathematics, though by the 1930s, interest in intuitionism began to wane (Hesseling 2003). Despite this, several key mathematicians continued to develop and refine Brouwer's ideas, keeping the program alive long after it had fallen out of favor. Even today, aspects of intuitionism remain a topic of discussion in fields such as philosophy, computer science, and physics (Posy 2020; Gisin 2019; Mark van Atten 2017). This chapter also seeks to understand why Brouwer's ideas were compelling enough to keep these mathematicians engaged, despite the wider community moving on. To do this, we will look closely at the stories of two influential figures—Abraham Fraenkel and Hermann Weyl—who found elements of Brouwer's vision compelling enough to pursue further. This chapter invites the readers to reexamine Brouwer's intuitionistic program not just as a failed scientific revolution, but as a deeply philosophical endeavor, one that raises essential questions about the very nature of how mathematics should be understood and practiced.

Synthese 200. https://doi.org/10.1007/s11229-022-03908-3

3.1 Building on the Giants: Friedman and the Evolution of Scientific Thought

In his theory of scientific philosophy, Michael Friedman builds on the ideas of thinkers like Kant, Schlick, Reichenbach, Carnap, and Kuhn, each of whom has offered important insights into the role of philosophy in shaping scientific progress (Friedman 2000, 2002, 2010). Yet, Friedman contends that none of these views, when taken on their own, fully explains how philosophical work can lay the foundation for a major scientific transformation. In order to grasp the core of Friedman's concept of scientific philosophy, and to understand what he adopts from these thinkers and where he diverges, it is worth taking a closer look at their positions through the lens Friedman provides.

3.1.1 From Kant to Kuhn and Beyond

Friedman, through Kant's perspective, portrays philosophy as operating on a distinctly different level compared to the empirical sciences. While psychology, mathematics, and geometry fall under the category of first-level sciences, philosophy serves as a second-level discipline, allowing us to comprehend and represent the objects of these first-level sciences. Kant, as Friedman describes, saw philosophy as a "transcendental inquiry" into the very conditions that make our understanding of objects in space and time, such as those provided by mathematical natural science, possible (Friedman 2001, 9). Newton's physics, both empirically successful and mathematically rigorous, presented a conceptual challenge for Kant: How could this theory be understood as rational? Kant's answer, according to Friedman, lies in the idea that fundamental concepts like space, time, motion, and force are not abstractions drawn from experience, nor do they hint at a metaphysical reality behind phenomena. Rather, they are the very structures that shape human understanding (Friedman 2001, 11).

In Kant's view, the rational success of Newtonian physics was anchored in a-priori, mind-given truths. These truths, which were fundamental to geometry and mechanics, underpinned our ability to even attempt empirical generalizations, like Newton's argument for the law of universal gravitation. Without these a-priori structures in place, such scientific extrapolations wouldn't be possible. However, as non-Euclidean geometries emerged, Kant's notion of an unchangeable a-priori truth began to falter. If the mind could conceive of non-Euclidean space, then Kant's idea that Euclidean geometry is an innate truth embedded in our mental capacities was challenged (Friedman 2001, 30–31). This dilemma prompted Friedman to reject fixed, unrevisable scientific frameworks, proposing instead a dynamic, evolving conception of a-priori principles—principles that develop in tandem with advances in the mathematical and physical sciences.

Schlick and Reichenbach, whom Friedman credits with continuing the Kantian project, helped shape this modern view. Schlick, whom Friedman calls the "first professional scientific philosopher" (Friedman 2001, 12), believed that philosophy provides the foundations for all scientific inquiry and that philosophical theories are intimately tied to the evolution of the sciences. Reichenbach's work on relativity added to this shift, arguing that a-priori principles aren't fixed but are constitutive of scientific concepts and evolve with scientific progress. The theory of relativity, for Reichenbach, demonstrated how the a-priori could change from Newtonian to Einsteinian frameworks.

It is in Carnap's *Logical Syntax of Language* (Carnap 1934) that Friedman finds a compelling expression of the logical empiricists' new approach. Logical rules, according to Carnap, are constitutive of the concepts of "validity" and "correctness", thereby relative to the linguistic framework in which they operate. Once the framework is chosen, its logical rules are a priori rather than empirical. In the late 1920s, Carnap was deeply influenced by Hilbert's meta-mathematical program, and he extended Hilbert's ideas to philosophy itself, seeing philosophy as a meta-science concerned with the logical structure of scientific language. This marked a shift from Kant's view, presenting philosophy as a formal, logical discipline—one that worked within frameworks, not above them.

As promising as Carnap's picture of dynamic yet constitutive principles of knowledge appeared, Friedman points to a lacuna in Carnap's account. Carnap argued that a framework, as a whole, cannot be judged as true or false but only as being fruitful or conducive to its purpose since a higher-level framework does not exist. In Friedman's view, Carnap completely overlooked "the relativistic predicament arising in the wake of Kuhn's work on scientific revolutions" (Friedman 2001, 57). Friedman criticizes Carnap for his lack of engagement with Kuhn's theory despite being the one who commissioned it, and probably one of the first to read it, which distances Carnap's theory from accounting for Kant's original questions regarding the possibility of mathematics and the natural sciences.

Next, Friedman turns to Kuhn and calls him to task on his theory of paradigm shifts in science, as it fails to show how a framework transition can be rational (Friedman 2001, 99). According to Kuhn, the idea of incommensurability suggests that when scientists move from one framework to another, the new framework often appears unintelligible from the standpoint of the old one. This leaves a problem that Friedman finds particularly troubling: how can such a radical shift be rationally motivated? How can scientists move to a new way of thinking when the new framework doesn't even seem possible according to the old one?

Friedman offers a solution by drawing on a distinction from the philosopher Jürgen Habermas between two kinds of rationality: instrumental and communicative (Friedman 2001, 53–56). Instrumental rationality is the reasoning we use to pursue our own goals, making decisions based on what best serves our individual aims. Communicative rationality, on the other hand, involves engaging with others through dialogue, working together to reach mutual understanding or agreement. Kuhn, in Friedman's view, confuses these two forms of rationality. Kuhn's attempt to establish universal, permanent criteria for rationality stumbles because it relies too heavily on instrumental rationality, which is subjective and varies from person to person. Communicative rationality, Friedman argues, is what really explains how scientific communities adopt new frameworks. It's not about individual goals; it's about reaching a collective consensus.

Friedman's theory of scientific framework transitions builds on this idea of communal rational consent. He envisions the process of moving from one framework to another as being mediated by what he calls a "higher-level meta-framework," which bridges the old and the new (Friedman 2001, 105). In Friedman's view, new frameworks don't emerge out of thin air; they evolve from the old ones. The principles that guide scientific thought are refined and developed through a process of inter-framework mediation. This allows older frameworks to be absorbed as special cases within the new ones, so that scientific progress becomes a kind of continuum. As each new framework builds on the old, Friedman sees the evolution of scientific paradigms as a convergent process, continually improving in generality and adequacy (Friedman 2001, 119).

Yet, Friedman's account hasn't escaped criticism. Some philosophers argue that his depiction of scientific revolutions is too mild, too neat. Alan Richardson, for example, suggests that Friedman's portrayal of scientific transitions as orderly and rational misses the point of what Kuhn described as true revolutions in science. According to Richardson, Friedman's version of scientific change is more of a "kinder, gentler" revolution, lacking the radical disruption that Kuhn believed was at the heart of paradigm shifts (Richardson 2002, 262–63). Menachem Fisch offers another criticism, questioning Friedman's claim that the mere availability of new ideas is enough to make them viable options for scientists. Fisch argues that to explain the rationality of scientists adopting new ideas, we need to consider the motivations and reasons that lead them to take these ideas seriously in the first place (Fisch and Benbaji 2011; Fisch 2017, 96). Robert DiSalle takes a similar stance, emphasizing that philosophy's role isn't just to justify a new theory after the fact but to actually help construct the new theory itself (DiSalle 2010, 528). DiSalle argues that the process of building new frameworks arises from the internal problems of the old ones, through a kind of conceptual analysis that reveals hidden assumptions and difficulties. This self-critical aspect of rationality, DiSalle insists, is essential for understanding how scientific frameworks change.

Despite these critiques, Friedman's overall vision remains compelling. While his critics point out gaps in his account, the central idea—that philosophy plays a crucial role in scientific revolutions—still stands. Friedman argues that philosophy and science are deeply intertwined, with scientific ideas often evolving out of philosophical debates. He maintains that in philosophy, there is never a stable consensus; the field is marked by constant shifts in thought, with no lasting agreement (Friedman 2001, 21–22). But this doesn't mean that science and philosophy operate independently of one another. On the contrary, Friedman believes that periods of consensus in science are times when practitioners operate within an agreed-upon framework, a shared set of norms and rules. However, during times of scientific crisis, when revolutions are in motion, the framework itself comes into question. It is then that philosophy provides a source of new ideas that can lead to the creation of new frameworks (Friedman 2001, 23).

Friedman's notion of scientific philosophy depicts philosophical arguments as entangled with and indispensable to revolutionary innovations in science. There is a special role Friedman assigns to philosophy as a meta-framework where questions about the normative boundaries of the scientific discipline can arise: questions that cannot arise from within the discipline itself, and that create the ground for new ideas to emerge and novel theories to develop. The exact dynamic relationship between scientific progress and philosophical ideas will be analyzed further in the next section, examining Friedman's example of the development of Einstein's relativity theory and the philosophical discussion that facilitated it.

3.1.2 Einstein, Helmholtz, and Poincaré: Transforming a Bold Idea into a Viable Theory

The story of Einstein's general theory of relativity, as told by Friedman, illustrates how philosophical developments can lay the groundwork for replacing an existing scientific framework with a radically new one. According to Friedman (2002, 193–95, 202–7), it was the philosophical debates on the foundations of geometry, particularly between Helmholtz and Poincaré, that gave Einstein the push to consider an idea that, from a Newtonian perspective, was neither mathematically nor physically possible—namely, that gravity may be represented by non-Euclidean geometry.[2] These philosophical discussions made it a "live option" for Einstein to rethink the relationship between gravity and the fabric of space-time.

For Newtonians, Einstein's general theory of relativity seemed unimaginable. The necessary mathematical tools hadn't even been developed until the late nineteenth century, and even after they were, the leap to connect them to physical reality was far from obvious. It wasn't until Einstein introduced the "principle of equivalence" that his new theory became physically plausible. But how could Einstein have believed that an abstract mathematical structure, like a four-dimensional semi-Riemannian manifold, could represent something as concrete as gravity? According to Friedman, this leap required not just scientific genius but a significant expansion of the intellectual landscape, one that had to be justified rationally (Friedman 2010, 499).

[2] There are differing opinions regarding the extent to which Poincaré and Helmholtz influenced Einstein's theory of relativity. John D. Norton contends that there is limited historical evidence to support Friedman's narrative, noting that Einstein himself credited Hume and Mach as key influences on his development of special relativity (Norton 2010, 2020). With this debate in mind, it is important to recognize that the account of Friedman's Einstein is presented here as a way to demonstrate Friedman's theoretical model in action. The focus is on evaluating the model itself, rather than verifying the historical accuracy of Friedman's claims.

Friedman suggests that to understand Einstein's breakthrough, we must consider the philosophical context of his time. Kant had previously provided a philosophical foundation for Newtonian physics, and over time, this foundation was modified and refined by figures like Helmholtz and Poincaré. Einstein, steeped in the ongoing debate over the nature of space, time, and motion, was able to connect the philosophical discussions about geometry with the empirical problems in physics. In doing so, he created a new geometrical framework that redefined our understanding of gravity and space-time itself.

Einstein's theory, while incommensurable with Newtonian physics, didn't emerge out of nowhere. It was rooted in the same philosophical tradition that had shaped Newtonian mechanics. The long history of debates over concepts like absolute and relative motion, dating back to Descartes, Huygens, and Leibniz, and carried forward by thinkers like Mach and Euler, helped set the stage for Einstein's new vision. According to Friedman, it was this grounding in an established philosophical tradition that made Einstein's radical framework transition rationally possible. It wasn't that Einstein's theory had to be immediately accepted, but that it could be seen as a *viable alternative* for practitioners familiar with Newtonian mechanics. As Friedman puts it, Einstein's theory had to be rationally considered by classical physicists—ignoring it would have been "irrational, unreasonable, and irresponsible," even if they ultimately rejected it (Friedman 2001, 108).

This notion of a rationally viable alternative brings us to Brouwer and his attempt to reform the foundations of mathematics. Like Einstein, Brouwer faced a significant philosophical challenge within his field—namely, the antinomies of Cantor's set theory, which raised serious questions about the foundations of mathematics. Brouwer's intuitionism provided a way to avoid these paradoxes by fundamentally changing how mathematical objects were conceived (Iemhoff 2020; van Dalen 1995). For Brouwer, mathematical entities weren't abstract symbols on a page, but mental constructions, created through human thought. His solution to the foundational problems in mathematics required nothing less than a complete overhaul of the discipline, a radical reform that would force mathematicians to give up widely accepted concepts and approaches (Brouwer 1907, 1912; Dummett 1977; Troelstra and van Dalen 1988).

In light of Friedman's model, we must ask whether Brouwer's intuitionism could trace its philosophical roots to an existing tradition, in the way that Einstein's theory could. If such a connection existed, how deep was it? Was Brouwer's program "rationally unacceptable," or could it be considered a "live option" for mathematicians at the time? In the upcoming sections, we will explore the philosophical roots of Brouwer's intuitionism in greater depth, assessing whether it aligns with Friedman's model of rational scientific transitions.

3.2 Tracing the Philosophical Lineage of Brouwer's Intuitionism

3.2.1 Kant

Brouwer's intuitionistic ideas can be traced back to Kant's philosophy, particularly to the concept of the a priori. A key piece of evidence for this connection is a 1906 correspondence between Brouwer and his thesis supervisor, Diederik Korteweg, where it becomes clear that Brouwer had studied Kant's *Critique of Pure Reason* and that it shaped his approach to mathematics.[3] In fact, Brouwer believed that at least part of his philosophical outlook on the a priori could be rooted in Kant's ideas, though he made significant adjustments to fit his own evolving perspective (van Dalen 1978; van Stigt 1990).

In the second chapter of his dissertation, Brouwer explores the a-priority of space and time, specifically affirming Kant's a-priority of time. According to Brouwer, apriority could mean one of two things: either (1) existence independent of experience, or (2) a necessary condition for the possibility of science. If we adopt the first interpretation, Brouwer argues, all mathematics—including non-Euclidean and Euclidean geometry, as well as both metrical and projective geometries—is a priori. If we consider the second alternative, he explains, where scientific experience arises from applying intuitive mathematics to reality, we can say that the only thing common to all mathematics, and sufficient to construct it, is the intuition of many-oneness, or what he calls the basic intuition of mathematics. Since this intuition leads us to perceive time as change itself, Brouwer concludes: "The only a priori element in science is time." (Brouwer 1907, 99).

Later in the same chapter, Brouwer addresses Kant's concept of space directly, stating that the main goal of his dissertation is to update Kant's view on apriority in experience (Brouwer 1907, 114). He refers to Karl Kehrbach's edited version of Kant's *Critique of Pure Reason* and explains that Kant viewed the perception of an external world through a three-dimensional Euclidean space as a fixed characteristic of human intellect. For Kant, imagining a different spatial perception for the same humans would be inherently contradictory.

Brouwer offers a distinct interpretation of Kant's argument about space and geometry. He suggests that Kant's reasoning hinges on two key assumptions: first, that all our external experiences occur within the realm of empirical space, and second, that this space adheres to the rules of three-dimensional Euclidean geometry. From these premises, Brouwer reasons that the properties of Euclidean

[3] The actual correspondence is translated in (van Dalen 1978, 298–99) and goes along these lines:

"Korteweg: Can one mention the name of Kant in a mathematical article?

Brouwer: Yes, Russell and Couturat did so, and the subject forces me to do so.

Korteweg: Did you study Kant sufficiently thoroughly to form an opinion?

Brouwer: I cannot prove that I did, but I have read *Die Kritik der Reinen Vernunft* repeatedly and seriously studied the passages I need."

geometry must be synthetic a priori judgments—truths that are both necessary and universal for any external experience.

But Brouwer doesn't stop there. He challenges this conclusion, arguing that our experiences can occur independently of any mathematical framework, including spatial conceptions. In other words, we don't need mathematics to structure our experiences. Instead, the act of organizing experiences into mathematical categories (such as formulating a concept of space) is, according to Brouwer, a free intellectual activity. We can choose to apply this framework or decide to experience reality without it. For him, this means that space cannot be considered a part of the a priori in the same way that Kant had envisioned.

This perspective is reinforced in Brouwer's Cambridge lectures, where he forcefully advocates for the separation of mathematics from theoretical logic (Brouwer 1981). He views intuitionistic mathematics as a purely mental activity, entirely detached from the constraints of language. It originates, as he explains, in the human perception of time itself. Brouwer vividly describes this perception as the "splitting" of a single life moment into two distinct parts, one following the other while still being remembered. This process of "twoity," when stripped of all specific content, forms the foundational intuition that underpins mathematics. In Brouwer's view, this abstract, time-based activity is the true source of mathematical thinking, far removed from the logical or linguistic structures we often associate with the discipline.

A natural question to consider is whether Brouwer's understanding of the intuition of time aligns closely with Kant's original idea (Mark van Atten 2004). However, rather than dwelling on whether Brouwer's reading of Kant was entirely accurate, it's more important to recognize how Brouwer himself perceived Kant's influence. For Brouwer, Kant's notion of the a-priority of time became a guiding light, deeply informing his own intuitionistic philosophy of mathematics. Brouwer elaborated on this connection by acknowledging that even though intuitionism had appeared weakened after certain developments in mathematics, it regained strength through a decisive move: rejecting Kant's idea of the a-priority of space while resolutely preserving the a-priority of time (Brouwer 1912). This refined view, which Brouwer called "neo-intuitionism," is centered on the splitting of life moments—each distinct and separate, but connected only through time itself. Brouwer saw this phenomenon as the essence of human intellectual activity, and the very foundation of mathematical thinking.

This basic insight, which he called the "intuition of two-oneness," not only generates the numbers one and two but also provides the foundation for all finite ordinal numbers (Brouwer 1912, 11–12). By repeating the process, one can generate higher numbers, even reaching the smallest infinite ordinal, ω. The same intuition of two-oneness also bridges the continuous and the discrete, ultimately leading to the concept of the linear continuum—a continuum that, unlike a collection of individual points, can never be fully broken down or divided into separate units.

This description forms the basis of what Brouwer called the "first act of intuitionism." Brouwer's first act was a bold move, one that separated mathematics from the confines of language and anchored it firmly in the flow of time (van Atten et al.

2002). In this act, Brouwer envisioned numbers being born from the process he called "twoity." It began with the simplest of experiences: the sensation of one moment passing into another, carried by memory. This act of holding onto one moment while moving to the next created a sense of unity—this was the number 1. Then, through repetition, the mind could build on this foundation. By adding one unity to another, the number 2 emerged, and then 3, and so on. In Brouwer's view, this mental process wasn't just a way of counting; it was the very origin of the natural numbers, rooted deeply in our intuition of time unfolding moment by moment.

Brouwer's perception of time gave him a unique way to distinguish between mathematical objects we can grasp intuitively and those that remain beyond our reach. Finite numbers and denumerable sets, he argued, are within our mental grasp—we can construct and understand them through our intuitive sense of time. But Cantor's collection of all real numbers, an infinite entity, was another matter entirely. For Brouwer, this kind of "actual infinity" was something we could never fully comprehend. Instead, he embraced the idea of "potential infinity," something that could be built step by step, never fully completed but always unfolding (Brouwer 1912, 1952). At the heart of Brouwer's view was an intuitive understanding of the continuum not as a collection of points strung together, as Cantor saw it, but as a whole, something that couldn't be reduced to its individual parts (Brouwer 1907, 8–9, 62). For Brouwer, this meant that Cantor's continuum, constructed from discrete points, was not a legitimate mathematical entity.

3.2.2 The French Semi-Intuitionists

Brouwer's philosophical influences extended far beyond the confines of Kant's ideas. The shaping of his intuitionistic thought was also deeply embedded in the French intuitionist tradition. Though Brouwer's contributions to mathematics were undeniably groundbreaking, his ideas—and even some of his biases—were reflective of the intellectual climate of his time. Despite his strong emphasis on independence and solitary reflection, Brouwer's distinctive philosophy and approach to mathematics emerged within a larger wave of European thought.

While Brouwer differed sharply from others on specific details, particularly regarding the roles of language and logic, he was firmly aligned with the intuitionist revival of the time, which he called 'Neo-Intuitionism.' Later, once his reputation as the father of mathematical intuitionism was well established, Brouwer looked back at this revival and referred to its members, such as Henri Poincaré, Emile Borel, and Henri Lebesgue, as the 'pre-intuitionist school.' By using the term 'Neo-Intuitionism,' Brouwer signaled his awareness of the long-standing tradition of intuitionist thought, a philosophical approach particularly strong in France. This tradition rested on two fundamental ideas: the dualistic separation of mind and matter, and the belief in intuition as a faculty that complements, rather than opposes, intelligence.

The pre-intuitionist school, later referred to as the French semi-intuitionists, mainly consisted of prominent figures like Rene Baire, Emile Borel, Henri Poincaré, and Henri Lebesgue. While these thinkers collectively influenced Brouwer's development of intuitionism, it's important to differentiate between Poincaré and the others, as their respective impacts on Brouwer's philosophical views on mathematics varied significantly. This section will focus on the contributions of Baire, Borel, and Lebesgue to Brouwer's thinking, while the next section will delve specifically into Poincaré's influence on the evolution of Brouwer's intuitionism.

Though Baire, Borel, and Lebesgue each held constructive ideas about the foundations of mathematics, none of them attempted to build a fully coherent philosophy of mathematics, leaving that task to philosophers and psychologists.[4] Among the three, Borel was the most vocal on foundational issues, articulating his belief that mathematical ideas could only achieve clarity through subjective mental reconstruction, rather than through rigid logical axioms or techniques (Bru et al. 2009). To Borel, mathematics was fundamentally a human mental activity expressed through language, and logical rules alone couldn't fully account for mathematical proofs. Logic, in Borel's view, provided a framework for mathematics but didn't constitute its core substance. As he eloquently put it, logic supplies the building material, but one must not confuse the mason with the architect (Borel 1907, 279).

Borel's perspective on mathematical existence was clear: consistency alone does not guarantee an object's existence in mathematics (Borel 1914). For him, there was another essential criterion—an object must also be effectively definable. In other words, it should be possible to describe or define the object using a finite number of words. If a number couldn't be defined this way, even if it was part of a consistent mathematical system, it remained unrealized, and therefore, in Borel's view, it did not truly exist. To illustrate this point, Borel provided an intriguing example: imagine a scenario where a denumerable number of people each choose a digit, one after the other (Borel 1928, 154). Although these choices could theoretically continue indefinitely, the number created through this process could never be defined in a finite number of words. As a result, this number, despite being logically conceivable, could not be considered "realized" in any meaningful mathematical sense. For Borel, such a number simply didn't exist. This example encapsulates his belief that effective definability is a crucial element of mathematical existence, setting a boundary for what can be considered real in mathematics.

However, Borel's approach to mathematical existence did not completely reject concepts like 'infinity.' Rather, he insisted that when such terms are used, they must be grounded in a well-defined law, expressed in a finite number of words, and accompanied by clear instructions on the sequence of steps involved. This notion of mathematical existence, as Borel described it, shares certain parallels with Brouwer's construction of choice sequences, a key component of Brouwer's intuitionistic

[4] Historians Dennis Hesseling and Alain Michel argue that the French semi-intuitionists never formed their ideas into a systematic philosophical stream and that they did not even refer to themselves as "intuitionists" but as "empiricists" or as "realists" (Hesseling 2003; Michel 2008, 160); it was Brouwer (1952) who was the first to refer to their theories as "pre-intuitionistic."

framework. Like Brouwer, Borel only recognized denumerable sets as legitimate, arguing that they alone could be clearly defined. This limitation led Borel to make a crucial distinction between two types of continua: the geometrical continuum, which he saw as naturally given and accessible through human intuition, and the numerical continuum, which, being uncountable and thus indefinable, he viewed as an artificial creation. For Borel, the latter lacked the definability required for true mathematical existence, reflecting a boundary between what can and cannot be understood or defined within mathematics.

The concept of intuition runs like a thread through Borel's work, though he never clearly defines its exact nature. For Borel, mathematical reality couldn't rest on logic alone; it had to be grounded in something more innate such as intuition (Borel 1898, 176–222). When it came to natural numbers and the set of natural numbers, Borel considered both to be "clear notions" for mathematicians, largely because there was a practical consensus among them in how these concepts were used (Borel 1914, 179). Where Brouwer and Borel part ways is in how they viewed the role of clarity in mathematical concepts. Borel believed that it was the responsibility of philosophers and psychologists to explore why certain notions appeared clearer to mathematicians than others. Brouwer, on the other hand, made these philosophical questions central to his intuitionistic theory, viewing them as crucial to understanding the very foundations of mathematics itself.

There's a clear and significant connection between the ideas of the French semi-intuitionists and Brouwer's development of intuitionism. Both camps shared a similar perspective on key issues like logic, mathematical proofs, and the very nature of mathematical existence. However, to understand the full story behind Brouwer's thinking, there's one figure who looms larger than the rest—Henri Poincaré.

3.2.3 *Poincaré*

Brouwer and Poincaré shared similar views regarding several critical issues in the philosophical debates surrounding the foundations of mathematics. Both thinkers challenged the idea that mathematics could be reduced to pure logic. For Poincaré, mathematical reasoning was far more than just a logical process as it held an inherent epistemic content that went beyond the boundaries of formal logic (Poincaré 1894, 371–72). He stressed that mathematics is synthetic in nature, rather than merely analytic. In his view, mathematics couldn't be constructed solely from logical principles. Poincaré believed that logic, on its own, could never derive general conclusions from specific instances, and thus mathematics was fundamentally irreducible to logic.

However, this wasn't to say that Poincaré dismissed logic altogether. Quite the contrary. He acknowledged the important role that logic played within mathematics, while also emphasizing the essential differences between the two. In his hands, logic retained its autonomy, but it couldn't account for the full complexity of mathematical reasoning. Poincaré saw logic as a part of mathematics, but not the whole.

His approach to mathematical logic was more flexible than Brouwer's intuitionism, which outright rejected classical logic. Yet, for Poincaré, the heart of mathematical reasoning wasn't about the logical connections between propositions—it was about the deeper epistemic relationships between judgments, underscoring a more dynamic understanding of mathematical thought.

Poincaré believed that the paradoxes of set theory, or antinomies, primarily emerged from the misuse of logic. In his view, logic was designed for finite sets, and applying it to infinite sets led to contradictions (Poincaré 1905). Brouwer, in a strikingly similar way, approached the issue with a firm rejection of classical principles (Brouwer 1919, 204). He argued that both the Solvability Axiom, sometimes called 'Hilbert's Dogma,' and the principle of the excluded middle were fundamentally flawed. According to Brouwer, these principles became accepted because classical logic was first derived from finite sets, and then, over time, it was mistakenly granted an a priori status. This misplaced universality led to the application of classical logic to infinite sets, where it no longer held true.

Though Brouwer and Poincaré are often seen as sharing a common ground in their views on the foundations of mathematics, the philosophical differences between them reveal a deeper divide, one that makes it difficult to draw a straight line from Poincaré's philosophy to Brouwer's intuitionistic theory. One of the starkest differences between the two lies in their attitudes toward logic and its role in mathematics (van Dalen 2012). For Brouwer, the paradoxes of set theory were not just technical glitches; they were red flags signaling fundamental problems with the way mathematics was built. Poincaré, on the other hand, saw them as minor flaws—issues that, while needing correction, didn't demand an overhaul of the mathematical foundations.

Poincaré believed that both intuition and logic were essential to mathematics, but they served different roles (Detlefsen 1993, 270). Logic, he argued, provided certainty—it was the tool used for proving things. Intuition, on the other hand, was the engine of invention, the means by which new mathematical ideas were discovered (Poincaré 1905, 37). While Brouwer agreed that intuition played a crucial role in guaranteeing the certainty of mathematics, he took it a step further, asserting that intuition was the very foundation of all mathematical construction. This is where their views diverge: for Poincaré, logic and intuition worked together, each with its distinct role, but for Brouwer, intuition was the bedrock, relegating logic to a secondary, supportive role.

Another key distinction between Poincaré's and Brouwer's philosophies lies in the role experience plays in their respective views of mathematics. For Poincaré, experience is essential; mathematical theories, in his view, are rooted in external experience (Heinzmann and Stump 2017). It is through this interaction with the external world that our minds are able to activate and apply their inherent capacities. Poincaré emphasizes that language plays a vital role in mathematical thought—not just as a convenient tool for communication or memorization, as Brouwer might argue, but as an integral part of the very process of conceptualizing mathematical ideas (Heinzmann and Nabonnand 2008, 177). For Poincaré, language is woven into the fabric of the experiential framework that underpins our cognitive abilities, serving as a bridge between our inner faculties and the outer world.

Brouwer, on the other hand, sees mathematics as a pure activity of exact thinking, completely independent of the external world (Brouwer 1907, 1912; Mark van Atten 2020). For Brouwer, mathematical truths are discovered through mental construction, and a proposition can only be considered true if the subject has experienced its truth through this inner construction. To "experience" something, in Brouwer's sense, is to engage in a mental act, wholly confined to the workings of the human mind. This stands in stark contrast to Poincaré's notion of experience, which involves interaction with the external world, enabling one to exercise the faculty of intuition.

The role of language is another point of significant disagreement between the two. Language, for Brouwer, was an afterthought—something that came into play only after the true mathematical activity, which, in his view, was deeply rooted in the pre-linguistic intuition of time (Brouwer 1907, 98–100). To Brouwer, mathematical thought was a silent, internal process, untouched by the external noise of words. Language, in his estimation, was merely a tool for communication after the real work had been done in the mind. For Poincaré, mathematics and language were inseparable. He believed that language was not only a means to express mathematical ideas but an active and essential component in the process of mathematical reasoning itself (Heinzmann and Stump 2017). Where Brouwer saw language as external and unnecessary, Poincaré embraced it as an intrinsic part of the cognitive process, necessary for shaping and clarifying mathematical thought.

Throughout this section, we've seen how elements of Poincaré's, Kant's, and the semi-intuitionists' philosophies surface in Brouwer's work. However, alongside these influences, there are also profound and unbridgeable differences between Brouwer's intuitionism and these other systems. One key divergence between Brouwer and Kant, for instance, lies in the very cornerstone of Brouwer's theory: the notion of intuition. Similarly, while Brouwer and Poincaré share common ground in their mathematical philosophies, they part ways on critical ideas such as the concept of experience and the role of language in mathematics. As for the semi-intuitionists, while their approach shows affinities with Brouwer's ideas, their contributions hardly constitute a unified philosophical tradition.

In contrast to the deep philosophical roots that Friedman outlines in Einstein's theory of relativity—where it is significantly shaped by the intellectual currents of Helmholtz and Poincaré—Brouwer's intuitionism doesn't exhibit such strong connections to Kant, Poincaré, or the semi-intuitionists. Moreover, Brouwer himself was insistent that his ideas were entirely his own. In a letter to Abraham Fraenkel, he expressed frustration at what he perceived as the "expropriation" of his work, claiming that certain German-speaking review journals had wrongly attributed his innovations to a collective tradition, implying that his contributions were shared with Poincaré, Kronecker, and Weyl. In Brouwer's view, his work was his "exclusive personal and spiritual property."[5]

When we examine Brouwer's reception through the lens of Friedman's theory, the mathematical community's reluctance to embrace his intuitionistic approach becomes more understandable. Unlike the classical physicist in Einstein's story, the classical mathematician did not seem to have the same kind of incentives to engage with Brouwer's ideas. Brouwer's intuitionism appeared as an isolated theory, disconnected from and, at times, directly contradictory to established mathematical practices.

Still, despite not gaining widespread acceptance within the mathematical community, Brouwer's intuitionism gained over 250 reactions between 1921 and 1933 (Hesseling 2003, 96) and attract the attention of prominent figures like Hermann Weyl and Abraham Fraenkel. What prompted these individuals to take Brouwer's ideas seriously? Why did they, unlike many of their peers, view intuitionism as a legitimate path forward? In the next section, we'll explore the intuitionistic episodes of Weyl and Fraenkel, examining the reasons behind their sympathetic responses to Brouwer's program.

3.3 The Appeal of Brouwer's Intuitionism

3.3.1 Abraham Fraenkel

Abraham Fraenkel's[6] path to engaging with Brouwer's intuitionistic ideas began through his deep interest in set theory. He had earned his Ph.D. in mathematics from the University of Marburg in 1914, where he also submitted his habilitation thesis a year later. At Marburg, Fraenkel initially focused on g-adic systems and the theory of abstract rings. However, after only a few years, he shifted away from rings and directed his attention toward set theory, a field that would shape his academic contributions (Corry 2004).

In 1919, Fraenkel published the first edition of his booklet, *Einleitung in Die Mengenlehre*, an introduction to set theory (Fraenkel 1919, 1967). At this point, Fraenkel's engagement with intuitionism was minimal—there were few references to intuitionistic ideas in this edition. However, by the time he published the second edition in 1923, Fraenkel had begun to grapple more seriously with Brouwer's intuitionistic philosophy. This new edition featured an extensive discussion of concepts central to intuitionism, such as mathematical existence and the principle of excluded middle, marking Fraenkel's growing interest in and engagement with intuitionistic ideas.

[5] As noted in a letter from Brouwer to Fraenkel, 28 January 1927 (van Dalen 2013, 390).

[6] Born as Adolf Fraenkel, and later in life, for obvious reasons, changed his name to Abraham A. Fraenkel.

Throughout the 1920s, Abraham Fraenkel solidified his reputation in the mathematical world through his work on set theory. His name became inextricably linked with "Zermelo-Fraenkel set theory," a refined version of Zermelo's axiomatic system, which Fraenkel had improved (van Dalen 2000). But what could make a mathematician so deeply embedded in the foundations of classical set theory reconsider those very foundations? And more intriguingly, why would he dedicate an entire section of the 1923 edition of his *Einleitung in Die Mengenlehre* to a careful examination of Brouwer's intuitionism, a theory that questioned the very core of set theory itself?

Fraenkel's engagement with Brouwer's ideas was not just a matter of academic curiosity but was made possible by a mix of personal and intellectual circumstances. His working knowledge of Dutch, stemming from his marriage to Wilhelmina Malka A. Prins from Baarn, made Brouwer's work more accessible to him. This linguistic advantage, combined with personal connections, paved the way for deeper engagement with Brouwer's intuitionism. During a visit to his in-laws in Amsterdam, Fraenkel took the opportunity to meet Brouwer (van Dalen 2013). What started as a single scheduled meeting quickly developed into a series of discussions, as the two mathematicians formed a collaborative relationship, exchanging comments on each other's work.[7] Brouwer even assisted Fraenkel in proofreading the second edition of his book on set theory in 1923. In turn, Fraenkel attended several of Brouwer's lectures at the University of Amsterdam and continued their conversations afterward. These encounters left a lasting impact on Fraenkel, influencing his approach to mathematical philosophy.

Fraenkel's genuine interest in Brouwer's intuitionism stood in stark contrast to the prevailing views of the German mathematical community at the time. In 1923, Fraenkel wrote to Brouwer, expressing his fascination with the "fresh life"[8] that intuitionism had brought into mathematics, noting that, despite being pronounced dead by many, these ideas were still actively fermenting in his own mind.

The exposition of Brouwer's intuitionism in Fraenkel's book on set theory can be described as courageous, given the mathematical community's unfavorable attitude, and especially in the light of Hilbert's criticism of Brouwer and Weyl (van Dalen 2000). Fraenkel made a point of including expositions of intuitionism in every book he wrote, thereby introducing intuitionism to a broader mathematical audience. Fraenkel's sympathetic expositions of intuitionism were a boost to Brouwer's

[7] In the course of their correspondence, Brouwer and Fraenkel experienced some friction when Brouwer became agitated by reading Fraenkel's sections about intuitionism in his *Ten Lectures on the Foundations of Set Theory* (Fraenkel 1927). It is unclear what exactly provoked Brouwer's outburst, as documented in Brouwer's letters, which came as a surprise to Fraenkel, who admired Brouwer and his work. Eventually, they made up and the correspondence between them continued. That same year Brouwer wrote Fraenkel that despite his stern remarks, "this letter is accompanied only by benevolent and friendly feelings towards you."(Brouwer to Fraenkel, 18 January 1927 (van Dalen 2013)).

[8] Letter from Fraenkel to Brouwer, 18/4/1923 (Hesseling 2003, 150).

efforts, who drew great encouragement from the fact that such a well-regarded mathematician was taking his views seriously.

Fifty years after the initial publication and eight years after Fraenkel's passing, a new edition of *Foundations of Set Theory* emerged, now co-authored with Yehoshua Bar-Hillel and Azriel Levy. This edition included an entire chapter dedicated to intuitionism, which was updated based on notes Fraenkel had left behind (Fraenkel et al. 1973). Fraenkel had always been diligent in ensuring that his books represented the fundamental ideas in mathematics, and this included Brouwer's intuitionism. He treated Brouwer's work not as a fleeting challenge to traditional mathematics, but as a serious contribution to the ongoing debate on the foundations of mathematics (van Dalen 2013, 391).

However, Fraenkel was not blind to the limitations Brouwer's intuitionism imposed on the field of mathematics. In a series of lectures on set theory delivered in Kiel in 1925, Fraenkel openly criticized intuitionism for the constraints it placed on analysis. He argued that intuitionism amputated a vital part of mathematics, likening this to the loss of a limb. Even after a body has healed, phantom pains may persist, and Fraenkel saw a parallel in the complex, almost painful nature of intuitionistic proofs and statements. They were, in his view, a "torso" of mathematics—missing crucial parts, making the process difficult to navigate (Fraenkel 1927, 58).

The correspondence between Fraenkel and Brouwer in the 1920s offers a glimpse into Fraenkel's inner reflections on the place of intuitionism in mathematics. Fraenkel understood the severe limitations that Brouwer's intuitionism imposed on the discipline but still believed that it deserved a voice within the broader mathematical landscape. However, his thoughts went beyond what was conveyed in his letters. In a notable section of the 1923 edition of his book, Fraenkel described Brouwer's intuitionism as both courageous and revolutionary, yet he remained unconvinced that it could serve as a viable alternative to the mathematical system he himself was committed to (Fraenkel 1923, 164). The reappearance of intuitionism in Fraenkel's posthumous works (Fraenkel 1967; Fraenkel et al. 1973) shows that he addressed intuitionism as part of mathematics long after the foundational debate had abated and the mathematical community's interest in Brouwer's theory decreased. Fraenkel was instrumental in disseminating Brouwer's ideas positively to the wider reaches of the mathematics community, though without recommending its endorsement.

Fraenkel's approach to intuitionism reflected his broader perspective on the philosophy of mathematics. He was a visionary thinker, someone who did not restrict himself solely to technical mathematical questions (Fraenkel and Cohen-Mansfield 2016, xix). While not formally a philosopher, Fraenkel embodied a philosophical spirit, one who could engage with and explain ideas that fell outside the mainstream, and his philosophical pursuits reached beyond the realm of mathematics.

In his 1930 essay, "Beliefs and Opinions in Light of the Natural Sciences," he touched on ideas that would later become central to Thomas Kuhn's theory of scientific revolutions—though Fraenkel arrived at them decades earlier (Fraenkel 1930). Fraenkel argued that science is not static, but constantly evolving. Yet unlike Kuhn, Fraenkel used the changing nature of scientific knowledge not to stress the limitations of science but to highlight the enduring relevance of philosophical and religious thought. He believed that, in the face of scientific uncertainty, philosophy of religion plays an essential role.

Fraenkel's inclusion of Brouwer's ideas, even when they ran counter to his own, speaks to his intellectual openness and belief in presenting a broad spectrum of thought to his readers. Fraenkel recognized that intuitionism held enough merit to warrant serious discussion within the mathematical discourse, and he wasn't alone in this consideration. Another eminent mathematician, Hermann Weyl, ventured even further into the realm of intuitionism, engaging with it so profoundly that, at one point, he seriously considered adopting it as a genuine alternative to classical mathematics.

3.3.2 *Hermann Weyl*

Hermann Weyl is often remembered as one of Brouwer's most significant converts to intuitionism, and even Brouwer himself viewed Weyl as his most radical follower (Beisswanger 1965; Bell 2000; Rosello 2012; Scholz 2004; van Stigt 1990). A prominent student of David Hilbert and an accomplished mathematician in his own right, Weyl had already made substantial contributions to mathematics before being drawn into the debate on its foundations. Like Hilbert, Weyl was deeply engaged with the philosophical questions surrounding mathematics. However, it was Weyl's unique interest in philosophy—particularly his engagement with Husserl's phenomenology and Fichte's metaphysical idealism—that ultimately inclined him toward Brouwer's intuitionistic approach (Scholz 2004; Sieroka 2009).

Weyl's path toward intuitionism, like that of Fraenkel, began with set theory, but his reasons for engaging with Brouwer were quite different. For Weyl, the paradoxes and antinomies of set theory were not just abstract philosophical puzzles to be set aside or solved in the margins—they were glaring signs of a deeper, more fundamental issue at the heart of mathematics (Weyl 1921, 86). In his view, these paradoxes revealed cracks in the very foundation of the discipline, and Brouwer's intuitionism seemed to offer a way to address these core problems head-on.

The summer of 1919 marked a pivotal moment for Hermann Weyl when he met Brouwer during a vacation in Engadin. Their encounter opened Weyl's eyes to the deeper complexities of constructive mathematics, and he almost immediately began incorporating Brouwer's intuitionistic notions into his own work, presenting intuitionism in a way that reflected his own perspective (van Dalen 2013, 299). By 1920 Weyl sent Brouwer a copy of his latest paper, titled "On the New Foundational Crisis in Mathematics" where he boldly declared that Brouwer's intuitionistic theory was the most viable alternative for rebuilding the foundations of mathematics on a solid, constructive basis (Weyl 1921).

Though Weyl enthusiastically supported Brouwer's ideas, he didn't ignore the practical difficulties they posed, and candidly acknowledged the restrictions of intuitionism on the everyday practice of mathematics (Weyl 1921, 109). Despite his initial excitement, Weyl's commitment to intuitionism wavered after just a few years, and he eventually distanced himself from Brouwer's approach, though he would return to it later in his career (van Dalen 1995).

Weyl's 1921 paper made a noticeable impact on the scientific community. Up until that point, discussions surrounding intuitionism had been limited to small, private conversations among a few key figures—Brouwer, Weyl, Fraenkel, and Hilbert. However, following the publication of Weyl's work, the interest in intuitionism expanded significantly. From 1921 onward, intuitionism became a central topic in the foundational debates, prompting mathematicians, logicians, and philosophers alike to respond and engage with the ideas. Weyl's endorsement of intuitionism in his paper played a vital role in bringing the theory into the mainstream of mathematical discourse.

That said, it's important to recognize that the version of intuitionism Weyl presented in his paper differed from the one Brouwer envisioned. As we saw in Chap. 2, while Weyl adopted key concepts from Brouwer—such as the concept of the developing sequence, his rejection of the principle of the excluded middle, and the idea of *functio continua*—he also introduced several distinctive ideas of his own, such as the role of universal and existential theorems and freely developing sequences (Weyl 1921, 109–10).

One of the mathematicians influenced by Weyl's turn to intuitionism was Fraenkel. The way Fraenkel expressed his ideas in *Introduction to Set Theory* and *Ten Lectures on the Foundation of Set Theory* suggests that he had absorbed at least some of Weyl's provocative metaphors. Fraenkel depicted mathematics as a kind of empire whose foundations were growing increasingly unstable, and intuitionism as an armed force ready to challenge it with "whetted arms" (Fraenkel 1923, 164). This imagery strongly echoed Weyl's own descriptions and made it clear that Weyl's influence had seeped into Fraenkel's thinking. On a deeper level, Fraenkel's reliance on Weyl was evident in his presentation of choice sequences, as he applied Weyl's 1921 paper as a foundation for explaining key aspects of the theory (van Dalen 2000).

The stories of Abraham Fraenkel and Hermann Weyl reveal how, when we shift our attention from abstract ideas to the lived experiences of individuals, framework transitions are bound to look very different. An individual's reasons to act in certain ways, to devote his efforts to considering certain theories in his published work while ignoring others, are a matter of personal exposure and social, political, and cultural circumstance. Weyl and Fraenkel, each in his own way and for his own reasons, seriously engaged Brouwer's intuitionism and produced colorful, clear, positive, and elaborate expositions of it. Those expositions contributed to the dissemination of Brouwer's ideas, thereby affecting the trajectory of the debate and the development of intuitionism during one of the most intriguing episodes of disagreement in the discipline of mathematics. According to Friedman, framework transitions are rooted in new ideas. But how new ideas spread, at least to some extent, owes to the place that they come to occupy in the deliberations individual practitioners conduct among themselves and others.

The environmental cost of transitioning to a new framework is another aspect of Friedman's theory that might shed new light on the lack of reception of Brouwer's intuitionism. Friedman argues that successive scientific revolutions resemble a series of nested developments and that earlier constitutive frameworks are exhibited as limiting cases of the new ones (Friedman 2001, 63). Hence, during framework transitions, the price practitioners committed to the old framework have to pay is relatively low. Following this line of thought, the disapproval of Brouwer's intuitionism did not occur merely because it lacked a solid philosophical tradition but also due to the enormous price intuitionistic mathematics coerced mathematicians to pay. Even the mathematicians who were willing to seriously reconsider their commitments to the old framework found themselves caught between a rock and a hard place; they were aware of the foundational problems that intuitionism seemed to solve, but they could not continue their scientific work without the mathematical theories that Brouwer's intuitionism forced them to renounce. Eventually, despite not having any concrete solution to the foundational problem, most practicing mathematicians chose to continue their everyday mathematical work and lived with the contradiction. To the community's eyes, the overall damage Brouwer's intuitionism caused to mathematics was greater than its achievements. This aspect of Friedman's theory provides yet another possible explanation of the reluctant response of the mathematical community to Brouwer's intuitionism.

3.4 Concluding Remarks

In Friedman's account of scientific progress, philosophical roots often serve as the foundation for rational shifts in scientific frameworks. This is evident in his analysis of Einstein, where established philosophical developments helped justify the transition to the theory of relativity. In contrast, Brouwer's intuitionism, with its more isolated and less clearly grounded philosophical origins, made it difficult for the broader mathematical community to see it as a natural or rational progression from the existing framework. Thus, while Brouwer's intuitionism may have posed solutions to significant foundational issues in mathematics, the absence of deep philosophical underpinnings left it vulnerable to rejection. In this sense, Friedman's model doesn't just explain why scientific ideas are accepted, but it also sheds light on why, despite its intellectual appeal, Brouwer's intuitionism was ultimately set aside. The mathematical community's response was not simply one of resistance to change—it was, in a way, a rational choice based on the perceived instability of intuitionism's philosophical grounding.

When viewed through Friedman's lens, Brouwer's intuitionism demonstrates that even unsuccessful revolutionary ideas—those not firmly anchored in a strong philosophical tradition—can still leave a lasting impact. They can travel through intellectual circles, and sometimes, even the most unconventional theories can change minds. The stories of Hermann Weyl and Abraham Fraenkel are prime examples of how these non-mainstream ideas captivated prominent mathematicians. But what exactly drew them to Brouwer's intuitionism? And, just as curiously, what prompted them to rethink their positions later? Friedman's analysis leaves these questions unanswered, though they are critical to understanding how intellectual shifts unfold through the deliberate choices of individuals. This topic will be further analyzed in Chap. 5, where we will explore Weyl's ongoing engagement with intuitionism. Before diving into that, however, we must turn our attention to Lukas Verburgt's post-Kuhnian historiography on probability theory, mathematics, and logic. As Verburgt's builds on Friedman's work and moves away from Corry's framework, the previous two chapters provided us the necessary background to properly engage with Verburgt's thesis.

References

Beisswanger, Peter. 1965. The phases in Hermann Weyl's assessment of mathematics. *Mathematic-Physical Semester Reports* 12:132–156.

Bell, John. 2000. Hermann Weyl on intuition and the continuum. *Philosophia Mathematica* 3 (8): 259–273.

Borel, Emile. 1898. *Lecons Sur La Theorie Des Fonctions*. 1st ed. Paris: Gauthier-Villars et fils.

Borel, Emile. 1907. La Logique et l'intuition En matMmatiques. *Revue de Metaphysique et de Morale* 15:273–283.

Borel, Emile. 1914. *Lecons Sur La Theorie Des Jonctions*. 2nd ed. Paris: Gauthier-Villars.

Borel, Emile. 1928. *Lecons Sur La Theorie Des Jonctions*. 3rd ed. Paris: Gauthier-Villars.

Brouwer, Luitzen Egbertus Jan. 1907. *On the foundations of mathematics*. Ph.D. Dissertation, University of Amsterdam.

Brouwer, Luitzen Egbertus Jan. 1912. Intuitionism and formalism. In *L.E.J Brouwer collected works I—Philosophy and foundations of mathematics*, ed. Arend Heyting, 123–138. Amsterdam: North-Holland.

Brouwer, Luitzen Egbertus Jan. 1919. Intuitionistische Mengenlehre. *Jahresbericht der Deutschen Mathematiker-Vereinigung* 28:203–208.

Brouwer, Luitzen Egbertus Jan. 1952. Historical background, principles and methods of intuitionism. In *L.E.J Brouwer collected works I – Philosophy and foundations of mathematics*, ed. Arend Heyting, 508–515. Amsterdam: North-Holland.

Brouwer, Luitzen Egbertus Jan. 1981. In *Brouwer's Cambridge lectures on intuitionism*, ed. Dirk van Dalen. Cambridge: Cambridge University Press.

Bru, Bernard, Marie-France Bru, and Kai Lai Chung. 2009. Borel and the St. Petersburg Martingale. *Electronic Journal for History of Probability and Statistics* 5 (1): 1–58.

Carnap, Rudolf. 1934. *The logical syntax of language*. London: Kegan Paul.

Corry, Leo. 2004. *David Hilbert and the axiomatization of physics (1898–1918): From Grundlagen Der Geometrie to Grundlagen Der Physik*. London: Springer.

Detlefsen, Michael. 1993. Logicism and the nature of mathematical reasoning. In *Russell and Analytical philosophy*, ed. A. Irvine and G. Wedekind, 265–292. Toronto: Toronto University Press.

DiSalle, Robert. 2010. Synthesis, and the synthetic a priori. And the origins of modern space-time theory. In *Discourse on a new method: Reinvigorating the marriage of history and philosophy of science*, ed. Mary Domski and Michael Dickson. Chicago: Open Court.

Dummett, Michael. 1977. *Elements of intuitionism*. Oxford: Clarendon Press.

Fisch, Menachem. 2017. *Creatively undecided: Toward a history and philosophy of scientific agency*. Chicago: The University of Chicago Press.

Fisch, Menachem, and Yitzhak Benbaji. 2011. *The view from within: Normativity and the limits of self-criticism*. Notre Dame: University of Notre Dame Press.

Fraenkel, Abraham. 1919. *Einleitung in Die Mengenlehre*. 1st ed. Berlin: Springer.

Fraenkel, Abraham. 1923. *Einleitung in Die Mengenlehre*. 2nd ed. Berlin: Springer.

Fraenkel, Abraham. 1927. *Ten lectures on the foundation of set theory*. Vieweg Teubner Verlag.

Fraenkel, Abraham. 1930. Beliefs and opinions in light of the natural sciences. *Hakirah, the Flatbush Journal of Jewish Law and Thought* 12:220–230.

Fraenkel, Abraham. 1967. *Lebenskreise. Aus Den Erinnerungen Eines Jiidischen Mathematikers*. Stuttgart: Deutsche Verlags-Anstalt.

Fraenkel, Abraham Adolf, and Jiska Cohen-Mansfield. 2016. *Recollections of a Jewish mathematician in Germany*. Springer.

Fraenkel, Abraham, Yehoshua Bar Hillel, and Azriel Levy. 1973. *Foundations of set theory*. Amsterdam: Elsevier.

Friedman, Michael. 2000. Geometry, construction, and intuition in Kant and his successors. In *Between logic and intuition: Essays in honor of Charles parsons*, ed. Gila Scher and Richard Tieszen. Cambridge: Cambridge University Press.

Friedman, Michael. 2001. *Dynamics of reason*. Stanford, CA: Center for the Study of Language and Information.

Friedman, Michael. 2002. Geometry as a branch of physics: Background and context for Einstein's 'geometry and experience'. In *Reading natural philosophy: Essays in the history and philosophy of science and mathematics presented to Howard stein on the occasion of his 70th birthday*, ed. D. Malament, 193–229. Chicago: Open Court.

Friedman, Michael. 2010. A post-Kuhnian approach to the history and philosophy of science. *The Monist* 93(4):497–517.

Gisin, Nicolas. 2019. Indeterminism in physics, classical chaos and Bohmian mechanics: Are real numbers really real? *Erkenntnis*. https://doi.org/10.1007/s10670-019-00165-8.

Heinzmann, Gerhard, and Philippe Nabonnand. 2008. Poincaré: Intuitionism, intuition, and convention. In *One hundred years of intuitionism (1907–2007)*, ed. M. van Atten, P. Boldini, M. Bourdeau, and G. Heinzmann, 163–177. Birkhäuser Basel: Publications of the Henri Poincaré Archives.

Heinzmann, Gerhard, and David Stump. 2017. Henri Poincaré. In *The Stanford Encyclopedia of philosophy*. https://plato.stanford.edu/archives/win2017/entries/poincare/. Stanford University.

Hesseling, Dennis E. 2003. *Gnomes in the fog: The reception of Brouwer's intuitionism in the 1920s*. Basel, Boston, Berlin: Birkhäuser Verlag.

Iemhoff, Rosalie. 2020. Intuitionism in the philosophy of mathematics. In *The Stanford Encyclopedia of philosophy (Winter 2019 Edition)*. Stanford University.

Michel, Alain. 2008. Remarks on the supposed French 'semi-' or 'pre-intuitionism'. In *One hundred years of intuitionism (1907–2007)*, ed. M. van Atten, P. Boldini, M. Bourdeau, and G. Heinzmann, 149–162. Birkhäuser Basel: Publications of the Henri Poincaré Archives.

Norton, John. 2010. How Hume and Mach helped Einstein find special relativity. In *Discourse on a new method: Reinvigorating the marriage of history and philosophy of science*, ed. Michael Dickson and Mary Domski, 359–386. Chicago: Open Court.

Norton, John. 2020. Philosophy in Einstein's science. In *Idealist alternatives to materialist philosophies of science*, ed. Philip MacEwen, 95–127. Leiden: Brill.

Poincaré, Henri. 1894. On the nature of mathematical reasoning. In *From Kant to Hilbert volume 2: A source book in the foundations of mathematics*, ed. William Ewald, 972–982. Oxford: Oxford University Press.

Poincaré, Henri. 1905. *The value of science*. Paris: Modern Library.

Posy, Carl. 2020. *Mathematical intuitionism. Elements in the philosophy of mathematics*. Cambridge: Cambridge University Press.

Richardson, Alan. 2002. Narrating the history of reason itself: Kuhn, Friedman, and a constitutive a priori for the 21st century. *Perspectives on Science* 10 (3): 253–274.

Rosello, Joan. 2012. *From foundations to philosophy of mathematics: An historical account of their development in the XX century and beyond*. Cambridge: Cambridge Scholars Publishing.

Scholz, Erhard. 2004. Philosophy as a cultural resource and medium of reflection for Hermann Weyl. https://arxiv.org/abs/math/0409596

Sieroka, Norman. 2009. Husserlian and Fichtean leanings: Weyl on Logicism, intuitionism, and formalism. *Philosophia Scientiae* 13 (2): 85–96.
Troelstra, Anne Sjerp, and Dirk van Dalen. 1988. *Constructivism in mathematics*. Amsterdam: North-Holland.
van Atten, Mark. 2004. *On Brouwer*. Belmont: Wadsworth/Thomson Learning.
van Atten, Mark. 2017. The development of intuitionistic logic. In *The Stanford Encyclopedia of philosophy (Winter 2017 Edition)*. Stanford University.
van Atten, Mark. 2020. Luitzen Egbertus Jan Brouwer. In *The Stanford Encyclopedia of philosophy*.. https://plato.stanford.edu/archives/win2017/entries/brouwer/. Stanford University.
van Atten, Mark, Dirk Van Dalen, and Richard Tieszen. 2002. Brouwer and Weyl: The phenomenology and mathematics of the intuitive Continuum. *Philosophia Mathematica* 10 (2): 203–226. https://doi.org/10.1093/philmat/10.2.203.
van Dalen, Dirk. 1978. Brouwer: The genesis of his intuitionism. *Dialectica* 32:291–303.
van Dalen, Dirk. 1995. Hermann Weyl's intuitionistic mathematics. *The Bulletin of Symbolic Logic* 1 (2): 145–169.
van Dalen, Dirk. 2000. Brouwer and Fraenkel on intuitionism. *The Bulletin of Symbolic Logic* 6 (3): 284–310.
van Dalen, Dirk. 2012. Poincaré and Brouwer on intuition and logic. *NAW* 5.
van Dalen, Dirk. 2013. *L.E.J. Brouwer—Topologist, intuitionist, philosopher: How mathematics is rooted in life*. London: Springer.
van Stigt, Walter. 1990. *Studies in the history and philosophy of mathematics: Brouwer's intuitionism*. Amsterdam: North-Holland.
Weyl, Hermann. 1921. On the new foundational crisis in mathematics. In *From Brouwer to Hilbert: The debate on the foundations of mathematics in the 1920s*, ed. Paolo Mancosu, 86–118. Oxford: Oxford University Press.

Chapter 4
Modernist Transformations: Rethinking the Origins of Change

Abstract This chapter analyzes Lukas M. Verburgt's historiographical framework for understanding transitions in the history of mathematics and scientific thought, particularly focusing on his concept of a 'break' between different intellectual frameworks. It examines how Verburgt's model reveals the incompatibility between two images of knowledge, even when they are rooted in the same a priori principles, making reconciliation difficult. The chapter then explores the impact of Brouwer's intuitionism on Soviet mathematicians such as Aleksandr Khinchin and Andrey Kolmogorov, showing how Brouwer's ideas prompted them to critically reassess their assumptions. The chapter concludes by questioning Verburgt's model, highlighting its limitations in explaining how framework transformation happen at the personal and communal level, and how individual mathematicians, motivated by social and political factors, either embraced or resisted new frameworks. It calls for a deeper exploration of these social and intellectual dynamics to fully understand such transitions.

Keywords Modernist transformations · Marxism · Mathematical modernism · Andrey Nikolaevich Kolmogorov · Aleksandr Iakovlevich Khinchin · A priori

Building on the previous chapters' exploration of Friedman's theory of the interplay between science and philosophy and Leo Corry's model of mathematics' layered structure, this chapter turns its attention to a new perspective. We dive into Lukas M. Verburgt's thesis, *A Terrible Piece of Bad Metaphysics,* which offers a post-Kuhnian historiography for understanding transitions in the history of scientific thought and integrates both Friedman's and Corry's views (Verburgt 2015).

Verburgt argues for the necessity of a fresh historiographical approach when examining the 'modernist transformation' in exact thought, and proposes that we must ask deeper questions about how certain ideas became thinkable within their intellectual environments. Central to his approach is the concept of a 'break' between different frameworks of thought (Verburgt 2015, 608). According to Verburgt, these breaks signify shifts in the foundational structures of knowledge—shifts that change what can be conceived and how intellectual developments unfold.

© The Author(s), under exclusive license to Springer Nature Switzerland AG 2025

K. Kish Bar-On, *Intuitionism Resocialized*, Studies in History and Philosophy of Science 66, https://doi.org/10.1007/978-3-032-05839-3_4

In this chapter, we explore this framework through the lens of Brouwer's intuitionism, situating Brouwer's philosophical and mathematical ideas within Verburgt's historiographical narrative.

As we move through the chapter, I also highlight key challenges that Verburgt's thesis raises, especially in relation to the Soviet mathematician Aleksandr Iakovlevich Khinchin's engagement with Brouwer's intuitionism. While Verburgt's analysis provides compelling insights into the broader intellectual context of Brouwer's work, I argue that his treatment of the 'breaks' between different a priori frameworks demands further refinement. This chapter thus serves as both an introduction to Verburgt's ideas and a critical examination of how his historiography applies to Brouwer's intuitionism, calling for a more detailed investigation of these intellectual ruptures.

4.1 From Non-Modern to Modern: Verburgt's Take on Framework Transitions

Verburgt opens his argument by addressing what he sees as a key issue in Leo Corry's model. Corry's model describes changes in mathematics on two levels: the body of knowledge and the image of knowledge. Corry contends that changes in the content of scientific knowledge—what he calls the "body of knowledge"—do not necessarily trigger shifts in the conceptual framework, or "image of knowledge," that underpins the discipline (Corry 2001, 177; 1989). To illustrate his point, Corry points to three well-known textbooks: Dickson's *Modern Algebraic Theories* (Dickson 1926), Hasse's *Höhere Algebra* (Hasse 1926), and Haupt's *Einführung in die Algebra (*Haupt 1929*)*. Corry further strengthens his argument by contrasting these with van der Waerden's *Moderne Algebra*, which did indeed signal a shift in the image of algebra, yet Corry insists that such changes in the "image" are not necessarily tied to changes in the "body" of knowledge (van der Waerden 1930).

Verburgt, however, takes issue with this stance. He argues that Corry's strict distinction between the body and image of mathematics oversimplifies the complex interaction between the two. Verburgt insists that these two dimensions of knowledge are far more intertwined than Corry allows, and he is particularly interested in demonstrating how shifts in the image of knowledge can have a transformative effect on the body of knowledge itself.

Verburgt offers four compelling examples of how the interaction between the "body" of knowledge and the "image" of mathematics led to transformative shifts from what he calls 'non-modern' to 'modern' conceptions of mathematics in areas like probability theory, algebra, and logic: first, he explains that certain non-modern ideas, such as viewing probability theory as part of natural science, were completely eradicated by the emergence of the modern image of mathematics. Second, he points out that some concepts that emerged from or were implied by the non-modern body of knowledge—such as the logic of classes, limits, probability axioms, algebraic freedom, and higher-dimensional geometries—became possible only through the lens of the modern image. Third, he highlights how modern ideas like measure-theoretic probabilities arose as a natural progression from non-modern developments. Lastly, he shows how the non-modern image inherently limited the body of knowledge by holding that probability, by its nature, couldn't be integrated into pure mathematics. The modern image, however, allowed for the incorporation of probability into the mathematical framework, expanding the body of knowledge in ways that had been previously unthinkable (Verburgt 2015, 11–12).

To construct a meaningful post-Kuhnian historiography for the study of probability theory, mathematics, and logic, Verburgt begins by engaging with two significant philosophical-historical perspectives. The first is the 'historical a priori' from historical epistemology, a concept championed by thinkers like Edmund Husserl, Georges Canguilhem, and Michel Foucault (Hyder 2003; Husserl 1939; Canguilhem 1967; Foucault 1966). The second is Michael Friedman's 'historicized a priori,' as developed in his "dynamics of reason" project, which reshapes our understanding of how scientific frameworks evolve over time (Friedman 2001; Richardson 2002).

By juxtaposing these two philosophical approaches, Verburgt embarks on developing his own philosophical-historical vision. His proposal suggests the existence of two distinct types of a priori: one rooted in the empirical world, where exact ideas are always related to empirical reality, and another in which exact thought remains entirely self-contained, without any outward reference to the empirical world (Verburgt 2015, 609). In the latter case, new exact ideas spring from the inner workings of the human mind, devoid of any direct relationship to the external world.

Verburgt's central thesis revolves around the idea that shifts in the realm of mathematics aren't merely changes from one 'image' of knowledge to another, while leaving the 'body' of knowledge intact. His prime example, the birth of modern probability theory, demonstrates how transformations between images of knowledge can actually dismantle concepts within the body of knowledge itself. The shift, Verburgt argues, was driven by the realization among key 'modern' thinkers—such as Boole, Hamilton, Hilbert, and Brouwer—that the source of a priori thought lies in the human mind. This realization was inconceivable to 'non-modern' thinkers like Venn, Gregory, and Von Mises, who believed that probability theory must always maintain a reference to the empirical world (Verburgt 2015, 13).

As a result of this new understanding, a transition occurred from the non-modern image of probability, which tied it firmly to empirical reality, to its modern image, where probability was seen as a construct originating in the human mind, devoid of external reference. Verburgt describes this shift as the 'de-philosophication of probability' (Verburgt 2015, 18). In the non-modern view, probability theory was deeply philosophical, grappling with fundamental questions about what probabilities really are. In the modern view, however, probability theory became part of mathematics, losing its philosophical foundation and being recast as a branch of measure theory. David Hilbert's call to axiomatize probability theory exemplifies this transition: by reducing probability to a set of mathematical axioms, Hilbert integrated probability into the broader mathematical framework. This effectively detached probability from its earlier philosophical roots, transforming it into a purely mathematical concept (Verburgt 2015, 19–20).

The attempt to reconcile the non-modern and modern images of probability theory under a single, unified body of knowledge is ultimately incomplete, since there's something fundamental missing from the modern body of probability theory that prevents this synthesis from fully succeeding. In other words, these two images of probability theory aren't just different perspectives of the same reality or a metaphorical shift in understanding; they are rooted in entirely different frameworks of thought. Each image corresponds to its own distinct a priori, a foundational set of ideas that shapes and governs how probability is conceived and applied within its respective framework.

In Verburgt's view, the transition from one image of knowledge to another involves more than a simple change in perspective—it introduces a 'break' in the body of knowledge itself, a gap that must be acknowledged and examined. This break, as Verburgt explains, marks a fundamental shift in the historical a priori, enabling the debates within modern mathematics to emerge (Verburgt 2015, 609). But what exactly happens during this 'break'? Is it a slow, evolving shift from one a priori to the next, or is it more like a sudden leap? How do new ideas spread within the uncertainty of a 'break'? And what about the cases where attempts to change the image or body of knowledge fail? To explore these questions more deeply, let us turn to the case of Brouwer's intuitionism, viewing it through the lens of Verburgt's framework.

4.2 Brouwer's Intuitionism: A Bold Transformation Attempt Between Images of Knowledge

Verburgt's exploration of Brouwer's intuitionism takes a slightly different route from the approaches encountered in the works of Fisch, Corry, and Friedman. While those scholars hardly discussed Brouwer or his intuitionistic endeavor, Verburgt directly engages with Brouwer's intuitionism and its role in the debate about the foundations of mathematics (Verburgt 2015, 469–526). He does so in a chapter devoted to Aleksandr Iakovlevich Khinchin and his contribution to the axiomatization of probability theory. However, he focuses exclusively on how Brouwer's intuitionistic ideas prompted Khinchin and fellow mathematician Andrey Nikolaevich Kolmogorov to critically examine the methods they used, and nowhere addresses the question of how Brouwer's intuitionism came into existence in the first place.

Towards the end of his dissertation, Verburgt begins to trace the origins of Brouwer's intuitionism, offering an intriguing glimpse into its philosophical foundations. He aligns his perspective with Friedman's argument from *Dynamics of Reason* (Friedman 2001), suggesting that key philosophical developments laid the groundwork for revolutionary mathematical innovations. Verburgt asserts that while both 'non-modern' thinkers and Kant himself may have found certain modern mathematical ideas inconceivable, it was the broader influence of neo-Kantian thinking that paved the way for these breakthroughs. He draws a parallel between the efforts to extend Kantian philosophy and the rise of new mathematical concepts. For instance, Hamilton's 'quaternions' emerged from an attempt to ground analysis in the notion of arithmetic as the science of pure intuition of time. Similarly, Brouwer's intuitionism was born from a reimagining of Kantian mathematical intuition, a distinct interpretation of Kant's ideas that allowed Brouwer to develop an entirely new mathematical framework (Verburgt 2015, 610).

Verburgt leaves us with only a hint of the story behind Brouwer's intuitionism. He briefly describes it as a 'modern' innovation, made possible by Kantian thought, yet he refrains from a deeper examination into the journey that led Brouwer to his intuitionistic vision. Brouwer's intuitionism presents a radical departure from classical mathematics, centering on the belief that mathematics is an entirely free mental activity, detached from both language and external objects. In Brouwer's view, mathematical truths are not universal and independent, but rather must be constructed in the mind to be understood as true. This perspective leads to a sweeping revision of concepts such as mathematical existence, where proof by contradiction—so integral to classical mathematics—is no longer valid. Fundamental principles, such as the law of excluded middle, are rejected, and with it, large portions of classical analysis are deemed unacceptable within an intuitionistic framework (van Atten 2020; van Dalen 2013; van Stigt 1990).

The implications of Brouwer's intuitionistic ideas on mathematics as a discipline affected both the mathematical body of knowledge and its image of knowledge. However, unlike Verburgt's story of probability theory, the transition suggested by Brouwer is not between two images corresponding to two different a prioris, but rather between two images corresponding to the same a priori. Intuitionism was rejected even by practitioners who adhered to the idea that the human mind is the source of a priori thought. The thing they were unwilling to accept was the consequences of intuitionism on the current body of mathematical knowledge. Brouwer's intuitionistic approach was not merely a restriction of classical reasoning; it contradicted classical mathematics in a fundamental way, making it incomparable with classical mathematics (Iemhoff 2020).

In Verburgt's story, the transition from one a priori to another shifted the philosophical concerns of probability theory from the image of knowledge to the body of knowledge. However, the fundamental structure of the field—the methods, proofs, and concepts—remained largely unchanged. In contrast, when we consider Brouwer's intuitionism through Verburgt's framework, we see that the shift from the classical image of mathematics to the intuitionistic image went beyond simple relocation. It required significant modifications to classical theories, with some concepts needing to be redefined to fit the intuitionistic framework, while other notions and methods had to be abandoned entirely. In a series of works known as *weak counterexamples*, Brouwer demonstrated that classical mathematical results—such as the intermediate value theorem, certain sequences of real numbers, and even Goldbach's conjecture—were not acceptable within his intuitionistic mathematics

(Brouwer 1928; Fletcher 2020). This made the shift far more radical and transformative than what we observe in Verburgt's story of probability theory.

Viewed through Verburgt's perspective, Brouwer's intuitionism reveals two distinct images of mathematical knowledge, both rooted in the same a priori but otherwise fundamentally different. These two images—the classical and the intuitionistic—approach the very foundation of mathematics in divergent ways, offering their own unique justifications for what constitutes legitimate mathematical knowledge. As a result, they give rise to entirely separate bodies of knowledge that share little to no common ground. Any attempt to reconcile these two images within the same mathematical framework becomes an impossible task.

4.3 Soviet Mathematicians and Brouwer: Khinchin, Kolmogorov, and Luzin's Intellectual Dialogue with Intuitionism

In the second part of his dissertation, Verburgt takes us on a journey into the heart of Soviet mathematical thought, where the tensions between formalism, intuitionism, and the quest for mathematical meaning came to a head. In a section titled "Moscow mathematics: formalism, intuitionism, and the search for mathematical content," Verburgt brings Aleksandr Iakovlevich Khinchin's perspective to the forefront (Verburgt 2015, 472). Khinchin, a key figure in Soviet mathematics, warned Marxist philosophers against hastily dismissing Brouwer's intuitionism as merely an "idealist" approach to the foundations of mathematics.

In 1926 Khinchin presented his paper, "Ideas of intuitionism and the struggle for a subject matter in contemporary mathematics", at a special colloquium on intuitionism at the Department for Natural and Exact Sciences at the Communist Academy. He opens the paper by stating that "we, Mathematicians, [...] are deeply convinced of the living reality of the subject which we study" and attempts to introduce Brouwer's intuitionism as a supportive view of such an approach (Khinchin 1926; Verburgt and Hoppe-Kondrikova 2016). As Khinchin saw it, intuitionism addressed mathematics as more than a formal logical play, but rather as having "real" references to nature as the subject matter of scientific inquiry (Vucinich 1999). Rejecting the intuitionistic theory amounts to rejecting the subject matter itself, so claims Khinchin, and concludes his defense of Brouwer's intuitionism in urging the readers

> to recognize that intuitionism has already refuted the intellectual perspective that prevailed in mathematics fifteen years ago. It has brought into the resources of the philosophy and the methodology of our discipline something that, perhaps, has not yet assumed its complete form and has not yet been fully comprehended, but that is clearly perceived by us as an inspiring and tangible achievement—one which promises to put us back onto the ground that we have almost lost from under our feet. (Khinchin 1926, 191; Verburgt and Hoppe-Kondrikova 2016)

Khinchin believed that framing intuitionism in a materialistic light would resonate with Marxist scholars, encouraging them to reconsider their reflexive, dogmatic rejection of philosophical ideas with idealistic origins. He recognized that Marxist philosophers might be quick to dismiss schools of thought like Brouwer's intuitionism due to its perceived alignment with idealism, but he was determined to show that intuitionism could be compatible with their worldview. In his 1926 paper, Khinchin skillfully portrayed Brouwer's intuitionism in a way that appealed to this Marxist audience. He even managed to offer a bold reinterpretation of intuitionism as "mathematical bolshevism" (Khinchin 1926, 84), cleverly connecting its revolutionary nature with the broader political upheaval of the time. This reinterpretation not only positioned intuitionism as a progressive force in mathematics but also as a philosophical approach that could align with Marxist ideals of radical transformation (Siegmund-Schultze 2004).

Beyond Khinchin's efforts, several other Soviet mathematicians were drawn to Brouwer's intuitionism. Among them was the prominent mathematician Andrey Nikolaevich Kolmogorov, who published a series of papers in 1925 in which he anticipated, to some extent, Heyting's formalization of intuitionistic logic and presented an elaborated account of the convoluted links between intuitionism, formalism, and other schools of thought (Kolmogorov 1925). According to Alexander Vucinich, Kolmogorov's work connected intuitionism with Soviet mathematical thinking, using it both to develop new ideas and to challenge the approaches of logicism and formalism (Vucinich 1999, 112).

Despite the Marxist theorists' continuous endeavors to present intuitionism as an idealistic philosophical stance that should be ignored, practicing soviet mathematicians maintain their initial endorsement of intuitionism's basic principles. Among those mathematicians was Nikolai Nikolaevich Luzin, Khinchin's doctoral supervisor and an influential member of the Soviet community of mathematicians. Luzin, who was often characterized as semi-intuitionist (Graham and Kantor 2009; Hesseling 2003), accepted the intuitionistic criticism of set theory's non-constructive existence proofs (Verburgt 2015, 485–87), and felt that intuitionism answered the fundamental questions that set theory alone could not answer (Vucinich 1999, 113).

Even though Luzin, Khinchin, and Kolmogorov viewed Brouwer's intuitionism favorably, none converted to intuitionism nor began to practice intuitionistic methods in their everyday work. According to Verburgt, Khinchin's defense of Brouwer's intuitionism was at least somewhat tactical (Verburgt 2015, 470), given the idealistic roots of Brouwer's theory. Overlooking Kant's influence, Khinchin intentionally simplified Hilbert's and Brouwer's positions on key foundational issues to portray Brouwer as the clear victor, framing his success as both evidence of the promise and the correctness of a Marxist philosophy of mathematics (Verburgt 2015, 483). Both Khinchin and Kolmogorov were willing to closely analyze the approaches and techniques they employed to achieve some of their most renowned results but neither embraced the intuitionistic point of view to the letter (Verburgt 2015, 470).

Several questions arise at this point: Why did Khinchin strive to make intuitionism look ideologically acceptable to his fellow members of the Communist Academy? What prompted Khinchin, leading probabilist and influential mathematician, to consider Brouwer's intuitionism in the first place, besides the immediate influence of his supervisor Luzin? Did Khinchin maintain his pro-intuitionistic views on the foundations of mathematics in later papers or renounced them altogether?

Verburgt ends the chapter with a brief, undecisive remark on Khinchin's motivation to endorse Brouwer's intuitionism:

> Given the superficiality of some of Khinchin's arguments this, of course, raises the question of what was tactical, what was real scientific interest and what was due to ignorance in his 1926 paper. Although this is, of course, far from a decisive answer, it seems to have been the case that it was all of these things at once; Khinchin wanted to caution his audience against a sweeping rejection both of intuitionism, as an orientation in the foundations of mathematics, and of more specialized intuitionistic studies without, perhaps, being fully aware of the sense in which these could conflict with his own set- and measure-theoretical studies. (Verburgt 2015, 499–500)

However, the rest of the picture remains unclear. Was Khinchin's support of Brouwer's intuitionism an episodic event or a continuous endeavor? Was it merely a fashionable act inspired by the general academic atmosphere in Russia during the 1920s, as implied by the words of Ioannis M. Vandoulakis: "even such a notable mathematician, as Aleksandr Jakovlevich Khinchin does not escape from a subjective exposition of Brouwer's ideas", while it was actually Kolmogorov whose "early works on intuitionistic logic contributed to its establishment as a mathematical discipline, rather than a philosophical doctrine" (Vandoulakis 2015, 156). Did Khinchin continue to discuss his foundational views regarding Brouwer's intuitionism in other works besides his 1926 paper? Verburgt mentions in a footnote that Khinchin corresponded with Arend Heyting in 1936 (Verburgt 2015, 476). What was the nature of their correspondence? Does it mean that Khinchin maintained his intuitionistic stance, at least until 1936?

To better understand Khinchin's willingness to re-examine his own methods in light of Brouwer's intuitionism, it is necessary to explore not only the act itself, namely, Khinchin's 1926 paper, but also Khinchin's commitments, motivations, and sources of influence in the broader context of his previous and later works. Whereas Verburgt presents a compelling account of the former, he leaves the latter out of the discussion.

4.4 Concluding Remarks: The Unexplored Depths of Modernist Transformations

Seen through Verburgt's lens, Brouwer's intuitionism takes on a deeper, more layered meaning. It doesn't just present an innovative mathematical theory, but it also raises compelling questions about Verburgt's concept of a 'break.' According to Verburgt, this 'break' serves as the divide between the 'non-modern' and the 'modern'—the critical moment where a new a priori replaces the old one. He suggests that this break marks the historical shift where the human mind begins to be accepted as the fundamental source for justifying and innovating knowledge within mathematics and logic. This moment in history, therefore, is not just a shift in mathematical technique but a profound philosophical turning point, where the intellectual foundation of mathematics itself changes dramatically.

If we were to plot the development of events along an imaginary timeline, we would place the 'non-modern' era first, corresponding to the a priori in which exact thought is connected to the external world. Then comes the 'break,' followed by the 'modern' era, which aligns with the new a priori, where the human mind becomes the source for all mathematical ideas. The 'break' itself, however, presents a unique enigma. It doesn't correspond to any a priori, because it represents the period of transition between two a prioris. But what actually happens during this transition stage, and how exactly does the transition take place? What role practicing mathematicians play in such transformations, and how they end up accepting a completely different viewpoint from the one they previously held? These questions are left unexplored in Verburgt's account.

Brouwer was driven by deep philosophical motivation to change the image of mathematics. He felt that the paradoxes of set theory are symptoms of a much profound foundational problem and that a massive reformation is required to achieve stability in the discipline and establish mathematics on solid foundations (van Stigt 1990). Building on this agenda, Brouwer suggested a new image of mathematics that, unlike the 'modern' image of probability theory, did not prevail. Viewed through the prism of Verburgt's work, it begs the question of whether the proposed 'gap' between the classical image and the Brouwerian image of mathematics can be described in terms of a 'break' a-la Verburgt.

Verburgt's scheme is significantly different from the three models discussed throughout this book. Unlike the models introduced by Corry, Friedman, and Fisch, Verburgt does not provide us with an account of framework transitions applicable to mathematics or all sciences; his argument on mathematical framework transitions is limited to transitions across the classical-modern divide. He discusses a particular kind of transition, from the non-modern to the modern, where at least one characteristic of the letter is a "formal axiomatic system" (Verburgt 2015, 10). The general direction of transition described in Verburgt's case studies drives away from the commonsense, intuitive aspects towards the formal and axiomatic aspects. Viewed through Verburgt's scheme, Brouwer's intuitionistic program took a step in the wrong direction in its attempt to replace formal theories ('modern', according to Verburgt) with intuitionistic ('non-modern') ones.

Verburgt is not alone in connecting modernism in mathematics to its emphasis on formal structures. Jeremy Gray defines mathematical modernism as a self-contained system of ideas that tends to focus on formal aspects while maintaining a complex, often uneasy relationship with the everyday world (Gray 2008, 1). However, as both Verburgt and Gray suggest, modernism in mathematics does not necessarily equate to progress. The push toward internal coherence and formalism can sometimes obscure the relationship between mathematical theories and the physical world, complicating their practical application. Brouwer's intuitionism serves as an interesting case in point if and when such an analysis will be realized, as it demonstrates a far-reaching attempt to separate progress from formalism.

Verburgt's concept of a 'break' illuminates the transition between non-modern and modern mathematics through changes in the a priori, but it leaves many questions unanswered. Why did this shift occur in the first place, and how did it so deeply affect mathematics? Verburgt does not explain the underlying motivations that drove mathematicians to adopt—or reject—the new a priori. Historically, mathematicians were not quick to embrace philosophical theories, especially if it required abandoning a successful working framework. What motivated some mathematicians, like Hilbert and Brouwer, to adopt this new a priori, while others, like Richard von Mises, resisted?

This brings attention to a gap in Verburgt's narrative—the personal motivations and social dynamics that influence such profound shifts. A change between a prioris is not instantaneous or automatic; it unfolds gradually, driven by the decisions and commitments of individuals operating within groups, communities, and institutions. These individuals are swayed by a variety of influences—personal, social, political, and philosophical—that may differ widely from one person to the next. This raises a new line of inquiry: how do individuals' interpersonal connections and internal reflections influence their decisions to adopt new frameworks? The next chapter will explore these dynamics by examining Hermann Weyl's wavering commitment to intuitionism, aided by Menachem Fisch's framework for understanding scientific transitions.

References

Brouwer, Luitzen Egbertus Jan. 1928. Intuitionistic reflections on formalism. In *From Brouwer to Hilbert: the debate on the foundations on mathematics in the 1920s*, ed. Paolo Mancosu, 40–44. Oxford University Press.

Canguilhem, Georges. 1967. *Mort de l'homme Ou Épuisement Du Cogito?* Editions de Minuit.

Corry, Leo. 1989. Linearity and reflexivity in the growth of mathematical knowledge. *Science in Context* 3 (2): 409–440.

Corry, Leo. 2001. Mathematical structures from Hilbert to Bourbaki: The evolution of an image of mathematics. In *Changing images in mathematics: From the French revolution to the new millennium*, ed. Umberto Bottazzini and Amy Dahan-Dalmédico, 167–186. Harwood Academic Publishers.

Dickson, Leonard Eugene. 1926. *Modern algebraic theories*. BH Sanborn & Company.

Fletcher, Peter. 2020. Brouwer's weak counterexamples and the creative subject: A critical survey. *Journal of Philosophical Logic* 49:1111–1157.

Foucault, Michel. 1966. *Les Mots et Les Choses*. Vol. 42. Paris: Gallimard.

Friedman, Michael. 2001. *Dynamics of reason*. Stanford, CA: Center for the Study of Language and Information.

Graham, Loren, and Jean-Michel Kantor. 2009. *Naming infinity: A true story of religious mysticism and mathematical creativity*. Cambridge: Harvard University Press.

Gray, Jeremy. 2008. *Plato's ghost: The modernist transformation of mathematics*. Princeton, NJ: Princeton University Press.

Hasse, Helmut. 1926. *Höhere Algebra*. 2 vols., Berlin: Sammlung Göschen.

Haupt, Otto. 1929. *Einführung in Die Algebra*. Vol. 1. Akademische Verlagsgesellschaft MBH.

Hesseling, Dennis E. 2003. *Gnomes in the fog: The reception of Brouwer's intuitionism in the 1920s*. Basel, Boston, Berlin: Birkhu-ser Verlag.

Husserl, Edmund. 1939. Die Frage Nach Dem Ursprung Der Geometric Als Intentional-Historisches Problem. *Revue Internationale de Philosophie* 1 (2): 203.

Hyder, David. 2003. Foucault, Cavaillès, and Husserl on the historical epistemology of the sciences. *Perspectives on Science* 11 (1): 107–129. https://doi.org/10.1162/106361403322286724.

Iemhoff, Rosalie. 2020. Intuitionism in the philosophy of mathematics. In *The Stanford Encyclopedia of Philosophy (Winter 2019 Edition)*. Stanford University.

Khinchin, A. Ya. 1926. Idei Intuitsionizma i Bor'ba Za Predmet v Sovremennoi Matematike. *Vestnik Kommunisticheskoi Akademii* 16:184–192.

Kolmogorov, A. N. 1925. On the principle of excluded middle. In *From Frege to Godel*, ed. Jean van Heijenoort, 414–437. Cambridge, MA: Harvard University Press.

Richardson, Alan. 2002. Narrating the history of reason itself: Kuhn, Friedman, and a constitutive a priori for the 21st century. *Perspectives on Science* 10 (3): 253–274.

Siegmund-Schultze, Reinhard. 2004. Mathematicians forced to philosophize: An introduction to Khinchin's paper on von Mises' theory of probability. *Science in Context* 17:373–390.

van Atten, Mark. 2020. Luitzen Egbertus Jan Brouwer. In *The Stanford Encyclopedia of philosophy*. Stanford University. https://plato.stanford.edu/archives/win2017/entries/brouwer/.

van Dalen, Dirk. 2013. *L.E.J. Brouwer—Topologist, intuitionist, philosopher: How mathematics is rooted in life*. London: Springer.

van der Waerden, Bartel Leender. 1930. *Moderne Algebra. Unter Benutzung Der Vorlesungen von E. Artin Und E. Noether*. 1sr ed. Berlin: Verlag Julius Springer.

van Stigt, Walter. 1990. *Studies in the history and philosophy of mathematics: Brouwer's intuitionism*. Amsterdam: North-Holland.

Vandoulakis, Ioannis M. 2015. On A.A. Markov's attitude towards Brouwer's intuitionism. *Philosophia Scientiæ* 19–1:143–158.

Verburgt, Lukas M. 2015. *"A terrible piece of bad metaphysics"? Towards a history of abstraction in nineteenth and early twentieth-century probability theory, mathematics and logic*. University of Amsterdam.

Verburgt, Lukas M., and Olga Hoppe-Kondrikova. 2016. On A.Ya. Khinchin's paper 'Ideas of intuitionism and the struggle for a subject matter in contemporary mathematics' (1926): A translation with introduction and commentary. *Historia Mathematica* 43:369–398.

Vucinich, Alexander. 1999. Mathematics and dialectics in the Soviet Union: The pre-Stalin period. *Historia Mathematica* 26:107–124.

Chapter 5
Navigating Ambivalence: Hermann Weyl's Path Through Intuitionistic Terrain

Abstract This chapter explores Hermann Weyl's intellectual journey, focusing on the personal struggles that shape scientific thought. By examining Weyl's evolving views on the continuum, logical existence, and foundational schools of thought, the chapter portrays his shifting positions not as simple indecisiveness but as part of a profound, continuous engagement with fundamental questions in mathematics. Applying Menachem Fisch's concept of normative ambivalence, it shows how external critiques led Weyl to challenge his deeply held commitments, driving his intellectual development. The chapter contrasts Weyl's persistent ambivalence with George Peacock's attempt at a hybrid solution in Treatise on Algebra, noting the different reactions each mathematician's work elicited within their community. Weyl's legacy, the chapter argues, lies not only in his mathematical contributions but in his rare openness to unresolved tension, embodying a thoughtful balance between philosophical inquiry and scientific rigor. Seen through Fisch's framework, Weyl's journey illustrates how individual ambivalence can catalyze broader shifts in understanding, underscoring the essential interplay between personal and collective progress in mathematics.

Keywords Normative ambivalence · Foundational crisis in mathematics · Paradigm shifts · Intuitionism · Constructivism · Formalism · Scientific framework transitions

As we step into this chapter, we shift focus from broad theoretical models to the more personal and nuanced journey of individual scientists. The previous models of Corry, Friedman, and Verburgt illuminated the complex processes behind framework transitions, but none of them have addressed the vital role of the individual scientist, their personal struggles, and their deep internal deliberations when confronted with groundbreaking new ideas. This chapter aims to fill that gap by delving into the life and intellectual journey of Hermann Weyl.[1]

[1] This chapter has been previously published as: Kish Bar-On, Kati. 2021. Towards a new philosophical perspective on Hermann Weyl's turn to intuitionism. *Science in Context*, 34(1): 51–68.

© The Author(s), under exclusive license to Springer Nature Switzerland AG 2025

K. Kish Bar-On, *Intuitionism Resocialized*, Studies in History and Philosophy of Science 66, https://doi.org/10.1007/978-3-032-05839-3_5

Weyl's turn to intuitionism presents a fascinating case study, not just of a mathematician wrestling with abstract ideas, but of a man caught in the tension between different schools of thought as he navigated a profound crisis in the foundations of mathematics. Through the exploration of three central themes—the nature of the continuum, the question of logical existence, and the necessity of intuitionism, constructivism, and formalism to adequately address the foundational crisis of mathematics—we are invited into his intellectual world. Here, we see not just abstract ideas at play but a deeply personal and ongoing endeavor to resolve some of the most fundamental problems in mathematics.

By following Weyl's shifting foundational positions, this chapter uncovers a new dimension in our understanding of scientific change. Drawing from Menachem Fisch's model of scientific framework transitions, the narrative here explores how moments of normative indecision, or ambivalence, played a crucial role in Weyl's contemplation of a radical shift toward intuitionism and away from it. Rather than seeing Weyl's shifting positions as indecisive, the analysis reveals them as the manifestations of a deeper, more intricate internal dialogue, one where he weighed new ideas, re-examined old assumptions, and responded to the evolving philosophical landscape of mathematics.

The exploration culminates in a comparison between Weyl's hybrid approach and that of George Peacock. Both mathematicians integrated various foundational ideas into their work, but their forms of hybridity led to distinct reactions from the mathematical community. Weyl's hybridity reflected his ongoing struggle, prompting more fluid and variable responses from the mathematical community, while Peacock's approach was more stable and elicited a different, more solid, kind of reaction. This comparison not only highlights the diversity of intellectual responses to foundational crises but also underscores how personal and communal dynamics influence the acceptance or resistance to new frameworks in mathematics.

5.1 The Intellectual Journey of Hermann Weyl

Hermann Weyl was born in 1885 in the small town of Elmshorn, located near Hamburg, Germany. His academic path took him to the University of Göttingen, where, from 1904 to 1908, he immersed himself in mathematics and physics, two fields that would define his life's work. Weyl's doctoral studies were conducted under Hilbert's supervision, and his dissertation, completed in 1908, explored integral equations, a subject closely tied to Hilbert's own groundbreaking research at the time. In the years following, Weyl made notable strides in the theory of self-adjoint operators, further establishing his mathematical prowess.

In 1912, Weyl delivered a lecture series on Riemann's theory of algebraic functions at Göttingen, which culminated in his influential book *The Concept of a*

https://doi.org/10.1017/S0269889722000047

Riemann Surface, published the following year (Weyl 1913). His career took a significant turn in 1913 when he secured a prestigious chair in mathematics at the Federal Institute of Technology in Zurich, where Albert Einstein was a fellow academic. It was here that Weyl's fascination with physics deepened, influenced profoundly by Einstein's revolutionary work, alongside the teachings of Hilbert, Felix Klein, and Hermann Minkowski.

Following Einstein's departure in 1914, Hermann Weyl began delivering lectures on the general theory of relativity, an endeavor that sparked his deeper engagement with both the mathematical and philosophical underpinnings of the theory. Three years later, in 1918, this focus bore fruit with the publication of two of his seminal works. *Raum-Zeit-Materie* (Space-Time-Matter) explored the nuances of relativity theory, offering a comprehensive mathematical treatment of Einstein's ideas (Weyl 1918a). Simultaneously, *Das Kontinuum* (The Continuum) emerged, where Weyl tackled the foundational crises in mathematics—particularly the emerging paradoxes within set theory—by attempting to construct a new basis for mathematical logic (Weyl 1918b).

The period from 1919 to 1926 marks a distinct phase in Hermann Weyl's intellectual journey, often referred to as his "intuitionistic phase." During the early 1920s, Weyl published a series of papers that actively defended the intuitionistic perspective in the foundational debate, showcasing his engagement with Brouwer's philosophical approach (Weyl 1921, 1925). Around the same time, Weyl began to explore the power and scope of group theory, which led him to develop a general theory of the representations and invariants of classical Lie groups (Weyl 1924, 1926a, b, c). His deepening interest in group theory coincided with the revolutionary developments in quantum mechanics, a connection that inspired him to apply his work on groups to this emerging field.

By 1928, Weyl published *Group Theory and Quantum Mechanics* (*Gruppentheorie und Quantenmechanik*), a book that made significant contributions by merging the theoretical advancements of group theory with practical applications in quantum mechanics (Weyl 1928, 1950). This work covered quantum theory, groups and their representations, along with applications involving rotation and permutation groups, as well as symmetric transformations. In 1930, Hermann Weyl received an invitation to succeed Hilbert at the prestigious University of Göttingen. It was a significant recognition of Weyl's stature in the mathematical community. However, the political situation in Germany soon began to deteriorate with the rise of the Nazi regime. As the political climate grew increasingly hostile, particularly toward academics of Jewish descent and liberal thinkers, Weyl faced a tough decision. Despite his strong ties to Göttingen, the escalating tensions and his wife's Jewish heritage made staying in Germany untenable.

In 1933, Weyl chose to leave and accept a position at the newly established Institute for Advanced Study in Princeton, a move that marked a turning point in his career. Once in Princeton, Weyl continued to make significant contributions in both mathematics and philosophy, though he gradually distanced himself from his earlier focus on physics. His intellectual pursuits during this period resulted in several important works, including *Mind and Nature* (Weyl 1934), which explored the relationship between the mental and the physical, and *The Classical Groups* (Weyl 1939), a fundamental text in the study of Lie groups and their representations. He also ventured into the theory of functions with *Meromorphic Functions and Analytic Curves* (Weyl 1944), and explored the aesthetic dimensions of science in *Symmetry* (Weyl 1952). His philosophical text *Philosophy of Mathematics and Natural Science* (Weyl 1949) reflects on the philosophical implications of scientific developments, continuing themes from his earlier work in 1927.

Weyl remained at Princeton until his retirement in 1951, after which he divided his time between Princeton and Zurich. Even in retirement, he continued to work, contributing to the field of number theory with his book *Algebraic Theory of Numbers* (Weyl 1954a). Tragically, Weyl's life was cut short by a sudden heart attack in 1955 while he was in Zurich, at the age of 70. His death marked the end of a remarkable career that bridged the worlds of mathematics and philosophy, leaving behind a lasting legacy.

5.2 From Constructivism to Intuitionism and Beyond

Hermann Weyl's journey into the foundations of mathematics began in 1910 when he started contemplating constructive methods as a serious alternative to classical mathematics (Beisswanger 1965; Scholz 2000). His evolving philosophical beliefs significantly shaped this direction. Kant's idea of the primacy of intuition played a foundational role in Weyl's intellectual development, pulling him towards constructive and later intuitionistic approaches (Scholz 2004; Sieroka 2009, 2019). Weyl cannot be classified as a strict idealist, since his metaphysical perspective was infused with elements of realism. However, during 1912, his philosophical ideas found new depth and direction through his engagement with Husserl's phenomenology and Fichte's metaphysical idealism (Bell and Korté 2015).

Brouwer, much like Weyl, was invested in the philosophy of mathematics and sought to advance Kant's a priori framework in mathematics (Brouwer 1907). Brouwer's philosophical direction was largely influenced by his mentor, Gerrit Mannoury, whose lectures presented mathematics not as an autonomous, independent domain but as a human creation deeply tied to the idealistic worldview that both Weyl and Brouwer would come to embrace (van Dalen 1999; Mannoury 1909).

However, despite the surface similarities between Weyl's and Brouwer's idealistic philosophies, their views diverged in meaningful ways. Brouwer's approach leaned strongly toward solipsism, a view where only the individual's mind is certain to exist, while Weyl held onto a phenomenological stance, claiming that objects and realities are perceived through human consciousness (Mancosu and Ryckman 2002). Nevertheless, their common philosophical ground, specifically their shared views about the primary intuition of time, contributed to Weyl's inclination towards intuitionism.

In 1918, with the publication of *Das Kontinuum*, Hermann Weyl's philosophical commitment to constructivism began to take shape (Weyl 1918b). This marked a significant turn in his thinking as he moved away from classical mathematics. However, less than three years later, his philosophical journey took another profound shift, and his constructivist approach evolved into a deeper engagement with intuitionism. This transformation was captured in his pivotal paper "On the New Foundational Crisis in Mathematics" published in 1921, where he begins by addressing the challenges inherent in Cantor's set theory (Weyl 1921). Cantor's set-theoretical foundations treated mathematical objects as abstract entities that existed independently of human thought. For Weyl, this notion was problematic, as he believed that mathematics was not about uncovering eternal truths in an abstract realm but about the structures created by human cognition (Feferman 1998). It was this very belief that led him toward Brouwer's intuitionism, which offered a framework based on the idea that mathematical objects are not independent entities but constructions of the human mind.

Nonetheless, during the 1920s, Weyl's allegiance to intuitionism began to waver, largely due to its inability to fully recapture the vast body of classical mathematics. Weyl increasingly realized that significant portions of mathematics, particularly those essential for physics, seemed to "dissolve in the fog" under an intuitionistic approach (Weyl 1931, 1949). By 1949, reflecting on this shift, he described the intuitionistic framework as "awkward" (Weyl 1949, 54) and even "intolerable" for practical mathematicians and physicists, emphasizing that mathematics should be in service to the natural sciences. In a correspondence with Oscar Becker,[2] Weyl confessed that while he admired Brouwer's philosophical rigor, intuitionistic mathematics was simply not what was needed in practice.

In the mid-1920s, Weyl began to distance himself from the intuitionistic program and moved toward Hilbert's axiomatic approach. Despite this move, Weyl never fully embraced Hilbert's program; he believed that modern mathematical research was a complex interplay between construction and axiom, rather than a rigid adherence to one approach. While he admitted the allure of choosing one perspective as the "genuine" way to think about mathematics, he also accepted the reality of their interdependence (Weyl 1940, 445).

Weyl's reflections on constructivism and axiomatics experienced at least two significant shifts. In the late 1930s, Weyl reaffirmed the relevance of his early constructive ideas, suggesting that they were critical to resolving foundational issues in mathematics (Feferman 1998). His internal debate between constructivism and axiomatics became even more pronounced in 1953, during a lecture titled "Axiomatic versus constructive procedures in mathematics" (Weyl 1954b). In this address, Weyl openly acknowledged the personal conflict he faced between these two approaches. He admitted that, while his heart was inclined towards constructivism, he felt a professional obligation to give precedence to axiomatics, believing that fairness required him to follow that direction despite his personal preferences. His final words in the lecture encapsulated this tension: "Indeed my own heart draws me to

---

[2] Weyl was quoted by Becker in a letter to Mahnke, August 22, 1926 (Mancosu and Ryckman 2002).

the side of constructivism… but justice seemed to require this from me" (Weyl 1954b, 38).

Weyl's frequent shifts in foundational thinking have been a subject of much discussion among historians and philosophers of mathematics. Solomon Feferman, for example, remarked that it is common for people to change their minds over time, and Weyl was no exception to this pattern (Feferman 2000, 181). Erhard Scholz, in his work, attributed these intellectual shifts to Weyl's philosophical roots, particularly his engagement with Fichte's ideas on space and matter (Scholz 2000). Similarly, Norman Sieroka analyzed Weyl's wavering stance on formalism, intuitionism, and constructivism as being deeply connected to his philosophical development, especially the distinction he made between Husserl's phenomenology and Fichtean constructivism (Sieroka 2009). However, such accounts only address part of the story, as they make no attempt to explain either Weyl's motivations for changing his mind or the roots of his undecidedness.

To do justice to Weyl's story, I will address the broader philosophical issues of normative framework transitions and try to explain how practitioners expound new alternatives. First, I will focus on three central themes that occupied Weyl's thought: the problematic differentiation between the intuitive and the mathematical continuum, the notion of logical existence, and the necessity of drawing upon intuitionism, constructivism, and formalism in order to adequately address the foundational crisis of mathematics. The analysis of these themes will reveal Weyl's continuous endeavor to deal with such fundamental problems. Afterwards, building on Menachem Fisch's model of scientific framework transitions and the special role that normative indecision or ambivalence plays in the process (Fisch 2017; Fisch and Benbaji 2011), I will examine Weyl's motives for considering such a radical shift in the first place, and I show that Weyl's constant changes of heart in regards to the foundations of mathematics were only symptoms of a deeper, intrasubjective process of self-criticism induced by exposure to external normative criticism.

5.3 The Undecided Nature of Weyl's Thought

5.3.1 The Concept of the Continuum

In *Das Kontinuum (Weyl* 1918b*)*, Weyl set out to reshape how we think about the very foundation of mathematics. Dissatisfied with the conventional set-theoretic approach to real numbers developed by Dedekind, Cantor, Weierstrass, and Hilbert, Weyl sought something different—something that could more closely align the abstract world of mathematics with the intuitive, continuous nature of our lived experiences. He was looking for a way to bridge the gap between the mathematical continuum and the one we feel and observe in everyday life: the flow of time, the movement of objects, the unfolding of space. His semi-constructive framework was an attempt to resolve this tension, a new lens through which the real number system *R* could be understood.

One of Weyl's frustrations lay in how the prevailing ideas of the time seemed to separate mathematics from the intuitive flow of reality. In the second chapter of *Das Kontinuum*, he used a deceptively simple example to illustrate this disconnect. Imagine a pencil lying on a table. It's there; we see it, we know it's there for the duration we observe it (Weyl 1918b, 87–88). For Weyl, this intuitive certainty carried great weight. However, he found it absurd that the rigorous logical principles used in mathematics could suggest otherwise—that mathematics could introduce new "moments" in time that our senses had missed, moments where the pencil might have been elsewhere. To him, this absurdity underscored the limitations of mathematical abstraction, particularly when it came to representing the continuous nature of time and space. He ended the second chapter of *Das Kontinuum* with a bold claim: the intuitive continuum of our experience and the mathematical continuum are fundamentally different. They do not, and cannot, coincide.

Less than two years after *Das Kontinuum*, Weyl's perspective on the continuum underwent a striking transformation. In his 1921 paper titled "On the New Foundational Crisis in Mathematics" (Weyl 1921), he distanced himself from the ideas he had previously defended. Weyl realized that his earlier belief, that the continuum was made up of individual, independent points, was fundamentally flawed. He saw that this view was rooted in the false assumption that each point in the continuum could stand alone, isolated from the rest. The new interpretation Weyl now embraced was Brouwer's intuitive construction of the continuum, as Weyl felt that Brouwer was the only one close enough to bridge the "chasm" between the intuitive and mathematical continua (Bell 2000).

In Brouwer's view, As Weyl saw it, a real number x wasn't a point on the number line, but rather an interval—a "dual interval" stretching between two numbers, y and z, with x contained somewhere within (Weyl 1921, 93). A real number, according to this new view, was not something fixed and static, but an entity that could be approximated infinitely. To Weyl, a real number was an infinite sequence of intervals, each one containing the next within it, each approximation narrowing the gap ever further, without ever fully pinning the number down. As Weyl described it in his 1921 paper, "a real number is an infinite sequence of dual intervals i, i′, i″, …, such that every interval of the sequence fully contains the next one inside it" (Weyl 1921, 93).

Weyl's reimagining of the real number system didn't stop with the introduction of dual intervals. He also brought two new concepts into play: sequences and laws. Both were used to define the relationship between the new individual real numbers and the new concept of the continuum, such that an individual real number is defined by a law ϕ that generates an infinite sequence, whereas the continuum is represented by a choice sequence whose development is unrestricted by any law and unfolds freely (Weyl 1921, 94).

A genuine continuum, according to Weyl, cannot be divided into separate fragments, since the true nature of the continuum demands that every part of it can be further divided without limitation (Weyl 1921, 115). The continuum in the traditional analysis was defined as a set of its points, the points being fundamental elements in the set. Each point was a part of the set, such that the traditional continuum could be divided into parts, but this was as far as one could go with such division since a point is considered a part that does not contain any further parts. By redefining primary elements of the construction of the real numbers as intervals rather than points, the mathematical continuum can finally be "in harmony" with the intuitive one. The chasm that represented the unbridgeable difference between the intuitive and the mathematical continuum in *Das Kontinuum* reappears now as the gulf between the new mathematical continuum and a set of discrete elements (Weyl 1921, 95).

Years later, in *Philosophy of Mathematics and Natural Sciences*, Weyl returned to the problem of the continuum, this time leaving it as an open question, still awaiting a full solution. He invoked the ancient philosopher Anaxagoras to illustrate his point: "Among the small there is no smallest, but always something smaller. For what is cannot cease to be no matter how far it is being subdivided" (Weyl 1949 [1927b], 41). This ancient insight, Weyl believed, perfectly captured the essence of the continuum—it is not a collection of discrete elements, as if nature had been "chopped off by a hatchet," but rather an unbroken flow, endlessly divisible.

Even as Weyl's career progressed, the mysteries of the continuum continued to haunt him. In his 1954 paper, "Axiomatic and Constructive Procedures in Mathematics," he returned once more to what he called the "riddle of the continuum" (Weyl 1954b). Despite decades of thought, Weyl regarded it as a puzzle still unsolved, a "serious affair" that mathematicians had yet to fully unravel.

5.3.2 Logical Existence

The first chapter of *Das Kontinuum* dives straight into the deep waters of logic and reasoning, introducing key concepts like "judgment," "category," and "proposition," and exploring how these ideas can be connected (Weyl 1918b, 5–9). To help illustrate complex logical ideas such as conjunctions ("and"), disjunctions ("or"), negations, and the concept of existence, Weyl devised six "complex judgment schemes" (Weyl 1918b, 10). One of his examples explains the logic behind being a father: If F(xy) means "x is the father of y," then F(Iy) means "I am the father of y." So, F(I∗) means "There is someone of whom I am the father"—in other words, "I am a father" (Weyl 1918b, 11). Judgement statements such as "every object has such and such a property" and "there is no object which lacks the relevant property" are deemed equivalent in *Das Kontinuum* (Weyl 1918b, 12). It is tacitly presupposed that a combination of existence judgement and negation (e.g., "there is no object that…") is fully interchangeable with universal statements such as "all" or "every" (Weyl 1918b, 12–13).

Fast forward a few years to Weyl's 1921 paper, *On the New Foundational Crisis in Mathematics*, and we find him proposing a very different theory. Building on his new understanding of the continuum as something that can be divided endlessly,

Weyl radically reshaped the concept of mathematical existence. Take the earlier example about fatherhood: in order to go from the statement "there is someone of whom I am the father" to the conclusion "I am a father," you need to produce evidence—a child, in this case. Without that concrete witness, the existence judgment "there is someone" loses its meaning.

Weyl took this a step further, criticizing the way logicians treat existential statements. He wrote that an existential statement like 'there exists an even number' is not, strictly speaking, a true judgment because it refers to no specific state of affairs and instead reflects an abstract generalization that is an empty construct devised by logicians, whereas a statement such as '2 is an even number' constitutes a genuine judgment grounded in an actual, concrete instance (Weyl 1921, 97–98). To Weyl, relying on existential statements without specific examples is like holding a piece of paper that says a treasure exists but doesn't tell you where to find it. The paper is worthless unless it's backed by an actual judgment (something like "2 is an even number") that reveals a concrete fact.

In *Philosophy of Mathematics and Natural Sciences* (1949 [1927]), Weyl made a sharp distinction between two types of logic: finite and transfinite. Finite logic deals with the familiar logical concepts of "and," "or," and "not." In contrast, transfinite logic incorporates existential judgments like "all" and "there is" (Weyl 1949, 13–14). Within the finite set, Weyl redefined logic as comprised of two logics and identified transfinite logic with existential and universal judgments. In order to deduce existence from an existence statement, it is no longer mandatory to provide a witness within a finite set, since one can eventually be found by pointing out each object. The same procedure applies for universal quantifiers since, within finite sets, each object can be accounted for. However, upon considering infinite sets, the latter no longer holds. Existence statements, as well as universal quantifiers, again receive the status of problematic at best and meaningless at worst. This approach to logic, specifically within the delimiter of finite and infinite sets, indicates that Weyl maintained his ideas from 1921 about logical existence and the problematic aspect of "there is" statements, but also that he accepted the validity of "there is" statements in some realms of logic (namely, those that address finite sets). This represents a more moderate approach than the one he presented in 1921.

In later works, the differentiation between finite and transfinite logic disappeared, and instead, Weyl used the terms finite and infinite logic to emphasize the differences he saw between two opposing mathematical views: one that consists primarily of construction, and the other that subsumes construction to axioms and deduction. The principles of classical, Aristotelian logic, Weyl argued, couldn't support existential and general propositions when applied to infinite fields, such as the sequence of integers or the continuum of points. The problem was particularly stark in the logic of the infinite. Even something as simple as the process of moving from one integer, n, to its successor, n′, couldn't be fully captured by axioms alone. As Kurt Gödel later demonstrated, there are always arithmetic truths that, while evident through constructive means, cannot be derived from any set of axioms, no matter how you frame them. At the same time, these axioms often overreach, making claims far beyond what can be justified by direct, constructive evidence. Hence,

Weyl concluded that when considering existential statements in infinite sets, it is only the "logic of the infinite" (i.e., Brouwer's intuitionistic logic) that comes closest to a true mathematical representation of our intuition (Weyl 1940, 446).

Despite his lifelong engagement with these ideas, Weyl never fully resolved the question of whether mathematics can truly reflect the nature of the world as we experience it. In his 1951 paper, Weyl admitted that the difficult issues he had identified, such as the logic of existential statements and the nature of the continuum, remained open questions (Weyl 1951). Neither the constructivist approach, with its focus on explicit construction, nor formalism, with its reliance on axioms, had managed to dissolve the tension between mathematical abstraction and the reality of human experience.

5.3.3 Being an Intuitionist, a Formalist-Finitist, or a Constructivist?

Beyond his famous work "On the New Foundational Crisis in Mathematics", Weyl revisited the topic of intuitionism two more times during the 1920s. The first was in 1924, when he presented a constructive proof of the fundamental theorem of algebra. The second was in 1928, during a seminar in Hamburg, when he found himself defending intuitionism against Hilbert's criticisms (van Dalen 1995). The Hamburg symposium note, in which Weyl stated that "in the epistemological evaluation of the new situation thus created, nothing separates me any longer now from Hilbert" (van Dalen 1995, 164), marks his gradual dissociation from intuitionism, which had already begun during 1926–1927 (Mancosu and Ryckman 2002).

Weyl had always been aware of the difficulties inherent in Brouwer's intuitionism. Even in his advocating 1921 paper, he acknowledged that while intuitionism offered a fresh perspective, it also imposed severe limitations. At first glance, the modest results of intuitionism might seem so limited that they could call the entire field of analysis into question. Weyl understood that this impression could lead to pessimism about the future of mathematics, but he reassured his readers that such pessimism was premature. He argued that while intuitionism's restrictions were real, they were not fatal, and that the next section of his work would show why these limitations could be overcome (Weyl 1921, 109). He firmly believed that mathematics, even in its logical forms, is ultimately tied to the natural numbers, and intuitionism provided a path to stay true to that foundation.

However, Weyl was also realistic about the practical implications of intuitionism for the broader mathematical landscape. On one hand, he praised Brouwer for bringing unparalleled clarity to the intuitive core of mathematics. Brouwer's approach, Weyl felt, allowed for the development of the early stages of analysis in a natural and intuitive way, keeping mathematics closely connected to human intuition. But on the other hand, when it came to more advanced and abstract mathematical theories, intuitionism created serious difficulties. The further mathematicians advanced, the more they ran into problems with the simple laws of classical logic, which no longer applied in these more complex realms. Weyl lamented this, noting that the inapplicability of classical logic led to "an almost unbearable awkwardness" in mathematics (Weyl 1949, 54).

As a whole, Brouwer's intuitionistic theory demanded a dramatic shift in thinking from anyone who embraced it. Followers had to give up some of the core ideas of traditional mathematics and replace them with completely different, and at times contradictory, concepts (Brouwer 1912, 1918; van Stigt 1990). Brouwer was willing to discard fundamental classical theorems and key principles in calculus, but for Weyl this was a step too far. While intrigued by intuitionistic ideas, he believed they had little impact on the actual practice of day-to-day mathematics and mathematical theorems.

Despite this, Weyl wasn't ready to fully embrace Hilbert's formalistic approach either. From his perspective, Hilbert's formalization of set theory reduced mathematics to little more than a collection of formal systems, stripped of any connection to their deeper, intuitive meanings. To Weyl, this felt like a step backward—a kind of degradation of mathematics. He described Hilbert's formalist approach as little more than an elegant game of formulas, perhaps even more entertaining than chess, but lacking any real connection to human understanding (Weyl 1949, 61). If the formulas were merely symbols with no inherent meaning, Weyl asked, what role did they play in the pursuit of knowledge? For Hilbert, the answer was simple: the focus of mathematics wasn't on the meaning behind the symbols but on the symbols themselves. For Weyl, this answer was not satisfactory.

As Weyl saw it, neither Hilbert's formalism[3] nor Brouwer's intuitionism could provide the ultimate solution for the foundation of mathematics. In his 1925 paper, "The Current Epistemological Situation in Mathematics", Weyl argued that both perspectives were crucial for gaining a comprehensive understanding of mathematical principles (Weyl 1925). He appreciated Brouwer's contribution in restoring the significance of intuition in mathematics, noting that Brouwer's focus on the intuitive grasp of concepts stripped away much of the abstraction, bringing a refreshing clarity to the core of mathematical thinking. Yet, Weyl also acknowledged that Hilbert's formalistic approach had its own indispensable value, particularly in addressing the human drive to represent ideas symbolically and grapple with concepts that go beyond immediate intuition (Weyl 1925, 141).

Even though Weyl never fully resolved the question of mathematics' ultimate foundations, he did see the possibility of a middle ground—a "dextrous blending"

[3] Brouwer and Weyl were the ones who described Hilbert's mathematical work as "formalistic," but that's not necessarily how Hilbert saw his own program. Hilbert's vision for mathematics aimed to formalize the entire field into a clear, axiomatic system and then prove that this system was logically consistent. The consistency proof, however, was meant to be achieved using what Hilbert called "finitary" methods, which he believed addressed the critiques raised by Brouwer and Weyl. So, while the term "Hilbert's formalism" is often associated with his approach—especially from the perspective of his critics—Hilbert himself likely saw his work as more aligned with finitism rather than strict formalism (Zach 2019).

of both approaches. This synthesis would combine Hilbert's formal axiomatic methods, which were particularly successful in fields like algebra (as championed later by Bourbaki), with constructive methods used in areas such as topology and geometry. This blend of approaches, Weyl believed, could offer a pragmatic solution for various branches of mathematics (Weyl 1954b).

Weyl did not regard intuitionism, constructivism, and formalism as clearly delineated theories. He noted that the borderlines of Brouwer's mathematics are vague, and some of Hilbert's mathematical considerations that appear to be evident, are not always entirely evident (Weyl 1953). Thus, he was not only concerned with questions regarding the subject matter of the theory itself (e.g., which definition of the continuum corresponds better with intuition?), but also with questions about what it means to be a formalist, a constructivist, or an intuitionist (e.g., What are the implications of building the foundations of mathematics solely on intuitionism\formalism\constructivism? Does such mathematics remain useful?).

Weyl is often depicted by historians and philosophers of mathematics as a "wanderer" (Scholz 2004, 1), someone who journeyed across various mathematical approaches and explored different philosophical terrains. His gradual retreat from Brouwer's intuitionism has been described as a process of "disillusionment" (Rosello 2012, 147). Solomon Feferman added to this picture, noting that Weyl's foundational views were constantly in flux, shifting over time (Feferman 2000). But Weyl's evolving positions on issues like logical existence, the continuum, and the nature of being an intuitionist, formalist, or constructivist, suggest something deeper than simple changes of heart. They reveal his persistent efforts to reconcile what he saw as the intuitive essence of mathematics with the practical, everyday applications of mathematical practice—without letting one overshadow the other.

In the remaining parts of this chapter, I wish to enhance the standard view held by many historians and philosophers of mathematics, and suggest that Weyl's indecisive acts are symptomatic of profound intrapersonal deliberations. By considering how normative transitions occur in general, and more specifically, by concentrating on the process of exposure to external criticism that is both normative and self-directed, the following sections aim to cast new light on Weyl's motivations and the reasons behind his hesitation.

5.4 The Power of Ambivalence: How Doubt Shapes Scientific Thought

In his recent work, Menachem Fisch has explored the intriguing question of how we can rationally change our normative commitments, even though, in principle, we might not recognize them as being flawed (Fisch 2017; Fisch and Benbaji 2011). Fisch introduces a novel twist to this dilemma: it isn't always through internal reflection that our norms begin to shift. Rather, it's often the influence of external criticism, especially when we find ourselves in the echo chambers of those who hold different commitments, that begins to destabilize us. This process, which Fisch describes as "an ambivalating process," can lead us to question norms we previously held with certainty (Fisch 2010, 530).

What Fisch reveals is that respected practitioners, when caught in this state of normative ambivalence, often create split or hybridic accounts of the foundations of their field in an effort to resolve their inner conflict. He points to historical examples like Tycho Brahe's attempt to merge a heliocentric planetary system with his essentially geocentric worldview, Galileo's inconsistent combination of Aristotelian and non-Aristotelian elements in his theories of free fall and projectile motion, and George Peacock's two-fold approach to algebra. Despite the confident tone of their later works, Brahe, Galileo, and Peacock were deeply committed to these split perspectives, a commitment born out of their inner struggles (Fisch and Benbaji 2011, 295–96). The ambiguous positions they took ultimately motivated other prominent figures in their respective fields to craft more coherent and balanced theories, moving the field forward in the process.

Fisch's concept of ambivalence shines a light on a fragile and often overlooked moment in the history of scientific revolutions—a moment that, if not carefully examined, can be mistaken for mere confusion or hesitation. Fisch argues that what makes certain historical figures so crucial during these transformative periods isn't just their groundbreaking discoveries, but the way their initial ambivalence is reflected in their attempts to grapple with conflicting ideas. In their efforts to reconcile opposing viewpoints, they often created hybrid solutions that seemed to split their field down the middle.

For historians, carefully dissecting these intellectual "splits" provides a window into the unsettled mindset of the thinkers behind them. But Fisch urges us to study these moments in a forward-looking way, as the scientists themselves experienced them—not as inevitable steps toward a more stable future but as acts of urgent, anxious engagement with their present dilemmas. Too often, we look back at these attempts as if they were just interim phases between one stable scientific framework and the next, but these moments are far more than unimaginative first tries. They are the raw, creative indecisions of people struggling to find their way through deeply unstable times, leaving behind traces of both brilliance and doubt (Fisch 2010, 536–37).

It's important to draw a clear distinction between making someone ambivalent and outright convincing them. To make a person ambivalent about their beliefs doesn't mean you've convinced them that those beliefs are wrong. In fact, as Fisch emphasizes, convincing someone that their deeply held normative standards are inadequate is virtually impossible. Instead, ambivalence creeps in when a person is exposed to external criticism—hearing other voices that question their norms. This criticism doesn't dismantle their belief system, but it plants doubt. Fisch describes this state of ambivalence as being caught between two minds (Fisch 2010, 537). On one hand, the individual begins to waver on certain aspects of their belief system, unsure of whether to hold on or let go. On the other hand, they remain fully committed to the rest of their normative framework, clinging to what hasn't yet been shaken.

This "ambivalated" state is a form of indecisiveness—not yet convinced that change is necessary but no longer fully confident in their original position.

Fisch's model of normative transformations offers two key insights into the concept of "becoming ambivalent," both of which are essential to understanding Weyl's intellectual journey. First, ambivalence is often misinterpreted as mere indecision or confusion. Second, ambivalence doesn't mean that someone has abandoned their old commitments or fully embraced new ones.

Different individuals respond to their ambivalence in different ways. Some manage to resolve it, making peace with one side or the other. But in many cases—including the historical examples Fisch gives—ambivalence remains unresolved, lingering in a person's work and thought, sometimes for their entire career. Weyl's case was much like this. His ambivalence toward different mathematical approaches never fully went away, and it continued to shape and complicate his work. In the following sections, we'll explore how Weyl confronted his own ambivalence, and how his struggle was perceived by mathematicians, philosophers, and historians of mathematics.

5.4.1 *Torn from Within: How Did Weyl Navigate His Own Ambivalence?*

Weyl's shifting views on the foundations of mathematics were deeply influenced by the intellectual environment he found himself in and the various voices he encountered along the way. One key factor was the critique of Cantor's set theory and Zermelo's axiom of choice by the French semi-intuitionists, who offered alternative perspectives (Hesseling 2003; Moore 1984). This external criticism left a mark on Weyl's foundational thinking, which became evident in his early work. In his 1910 paper on the definition of key mathematical concepts (such as definable relations) and later in his groundbreaking book *Das Kontinuum*, Weyl wrestled with these influences (Weyl 1910, 1918b). In the preface to *Das Kontinuum*, he expressed hope in what he saw as a promising "halfway" solution: a constructive approach that would steer clear of the paradoxes found in Russell's work and the troubling implications of Cantor's set theory.

Although Weyl wasn't a professional philosopher, philosophical ideas—particularly idealism—played a significant role in shaping his thoughts on mathematics. During his student days at Göttingen, he was drawn to Edmund Husserl's lectures, where he was exposed to the emerging field of phenomenology. Later, during his years at ETH Zurich, his friend and colleague Fritz Medicus, a professor of idealist philosophy, encouraged him to delve into the work of Johann Gottlieb Fichte (Scholz 2004; Sieroka 2007, 2009). These philosophical influences, especially from Husserl and Fichte, are evident in *Das Kontinuum* and Weyl even mentions Husserl by name in the book's preface, noting that on the epistemological side of logic, he found himself in agreement with Husserl's views (Weyl 1918b, 2).

This philosophical foundation provided Weyl with a framework for addressing one of his central concerns: the continuum problem. His work on this issue wasn't merely a technical investigation into mathematical concepts; it was also a philosophical endeavor aimed at bridging the gap between what we intuitively experience and the formal, mathematical structures we rely on to interpret the world. By grounding his approach in critical epistemology, Weyl sought to explore the deep connections between our immediate, intuitive perceptions and the abstract systems we use to model reality—whether in the realm of geometry, physics, or beyond.

At the start of *Das Kontinuum*, when discussing the nature of judgment and property, Weyl paid homage to Fichte's metaphysical philosophy, acknowledging his importance in shaping a clear understanding of what judgment truly means (Weyl 1918b, 7). However, much like his evolving views on the foundations of mathematics, Weyl's philosophical outlook also underwent significant changes over time. In a 1954 lecture titled *Insight and Reflection*, Weyl reflected on his long intellectual journey, which had begun with Kant's ideas, moved through idealist phenomenology, and eventually led him to recognize some serious issues with this approach (Weyl 1955). As Weyl grappled with these problems, his thinking shifted, drawing him closer to the ideas of philosophers like Ernst Cassirer in some respects, and even theologists like Meister Eckhart in others (Bell 2004). Over the years, even the strong influence of Fichte's ideas on his thinking began to fade, and by the end of his life his reflections pointed toward a philosophical position more aligned with Heidegger's existentialism or Gonseth's dialectical methods (Scholz 2004).

Henri Poincaré, too, played a pivotal role in shaping Weyl's foundational and philosophical views. The impact of Poincare's philosophy of mathematics, specifically his position regarding impredicative definitions and the concept of existence and potential infinity (Poincaré 1963), is present in Weyl's 1918 works (Weyl 1918a, b). In *Das Kontinuum*, Weyl accepted a significant part of Poincare's definitionist philosophy, including Poincare's conception of the natural number and the idea that there are no completed infinite totalities (Feferman 1998, 2000). Poincaré's predicativist approach led Weyl to revise some aspects of Husserl's semantics and ontology. When it came to defining natural numbers, Weyl's own definition became a unique blend of Poincaré's perspective and Husserl's phenomenological framework, reflecting his ability to synthesize the ideas of multiple thinkers (Feist 2002).

Two years after the release of *Das Kontinuum* and *Raum-Zeit-Materie*, Hermann Weyl's intellectual journey took an unexpected turn. In 1921, he published "On the New Foundational Crisis in Mathematics" (Weyl 1921), a work that signaled yet another shift in his thinking, one shaped by the ideas of a new influence, Brouwer. Weyl had been aware of Brouwer's work since at least the early 1910s, having referenced Brouwer's groundbreaking papers on topology in his own 1913 paper on Riemann surfaces (van Dalen 2013). There's even evidence that the two had met briefly in 1912 during Brouwer's visit to Göttingen, but it remains unclear whether they discussed foundational issues at that time.

The real turning point came in the summer of 1919, when the two met for the second time during a summer vacation in Engadin, and Weyl heard the story of Brouwer's intuitionism firsthand (van Dalen 1995; Hesseling 2003). Brouwer introduced Weyl to his intuitionist theory and challenged the very core of the formalist mathematics Weyl had long embraced. Over the course of their time together, Brouwer's critiques stirred something in Weyl, planting seeds of doubt about the foundations of mathematics he had once accepted without question. Deeply inspired by Brouwer's bold ideas, Weyl penned "On the New Foundational Crisis in Mathematics", a paper that marked a dramatic departure from his previous work. The constructivist structure Weyl had adhered to in *Das Kontinuum* was no longer enough. He now felt compelled to explore new paths, to face the very challenges Brouwer had raised, and to search for answers in the fertile ground of intuitionism.

By the time Hermann Weyl wrote the preface to the 1932 edition of *Das Kontinuum*, his earlier optimism about solving the foundational problems of mathematics through Brouwer's intuitionism had started to fade. He still saw Brouwer's original approach as an important step in the journey of foundational research, but he recognized that the landscape had changed. Two major trends—Intuitionism and Formalism—had come to dominate the field, and yet, despite the strides these movements had made, Weyl felt that neither had delivered a truly satisfying answer to the big questions. The foundations of mathematics, he admitted, remained unsettled, and the problem still felt as elusive as ever. Weyl candidly acknowledged that without a radical overhaul, the ideas in *Das Kontinuum* no longer aligned with his current thinking (Weyl 1918b, 2–3).

This sense of uncertainty stayed with Weyl, casting a shadow over much of his later work. In his 1946 review paper, he reflected on the crisis that had haunted mathematicians for nearly fifty years. "We are more uncertain than ever," he wrote, about the ultimate foundations of logic and mathematics (Weyl 1946, 11). Outwardly, this crisis hadn't stopped mathematicians from doing their daily work. But for Weyl, the uncertainty had taken a personal toll. It pushed him toward areas of research he considered relatively "safe" and dampened the enthusiasm that had once driven his mathematical pursuits. He suspected that many of his peers, mainly those who cared deeply about the philosophical meaning of their work, must have shared this struggle, feeling the tension between their scientific efforts and the larger, existential questions of human life: what it means to know, to suffer, to create, and to make sense of the world.

Even as Weyl's career neared its end, his intuitionistic leanings never fully disappeared. In one of his last papers, written in 1954, Weyl still expressed a deep sense of uncertainty about the very foundations he had spent a lifetime exploring. He openly admitted that the constructive approach to the continuum of real numbers remained, in his words, "a serious affair." He went so far as to say that, even after all these years, the logical issues involved were far from fully clarified or settled (Weyl 1954b, 17).

These reflections, taken from different stages of Weyl's long and illustrious career, offer a glimpse into his self-awareness and the ongoing internal dialogue he maintained about his shifting beliefs. In 1932, he acknowledged the need to reconcile his earlier views from 1918 with his newer perspectives. In 1946, he confessed that his lingering uncertainty had shaped the course of his scientific work. And by 1954, he referred to himself as "bold" for maintaining an undecided position even after so many years of contemplation. These are the words of a man fully conscious of his ambivalence—someone who was aware of both the weight of his indecision and the complexity of the issues he grappled with as a practitioner in the scientific community.

5.4.2 Confused from Without: How Do Weyl's Scholars Interpret His Ambivalence?

Mathematicians, philosophers, and historians have long wrestled with Hermann Weyl's ever-changing views on the foundations of mathematics. Solomon Feferman, for instance, saw Weyl's work in *Das Kontinuum* as a significant contribution to the predicativist program (Feferman 1988, 16). He believed that Weyl's stance on predicativity stemmed largely from his critique of set theory, particularly its use of vicious circles, and his firm belief in the irreducibility of the natural number (Feferman 1998). For Weyl, mathematical concepts like sets and functions had to be defined explicitly, without relying on the abstract, ambiguous methods of set theory. Even though Weyl repudiated his ideas from Das Kontinuum in favor of Brouwer's intuitionistic ideas only two years later, Feferman maintained that Weyl continued to regard the predicative approach as being of genuine value and that he "never really gave up the achievements of his 1918 monograph" (Feferman 2000, 181).

Norman Sieroka saw Weyl's shifting foundational views as closely tied to the evolution of his philosophical thinking. In *Das Kontinuum*, Weyl rejected the idea of reducing mathematics to logic and set theory, a stance influenced by both Fichte and Husserl. But by the 1920s, Weyl began to differentiate between the two thinkers, linking Husserl's phenomenology to Brouwer's intuitionism and Fichte's constructivism to formalism. After briefly embracing the "intuitionistic-phenomenological" approach, Weyl came to believe that Fichte's "formalistic-constructivist" path might hold the key (Sieroka 2009, 90). Sieroka found parallels between Weyl's interdisciplinary thinking, ranging from mathematics to physics and the theory of subjectivity, and argued that Weyl's exploration of these "intellectual neighborhoods" laid the groundwork for a deeper understanding of transformations and invariances, especially in the context of historiographical questions (Sieroka 2019, 120).

Biographical notes about Weyl written by mathematicians often dismiss his intuitionistic deliberations or ascribe them to his philosophical interests, which they consider to be entirely separate from his mathematical work. Michael Atiyah's short essay, for instance, analyzes Weyl's contributions to group theory and quantum mechanics while overtly ignoring his engagement with intuitionism and constructivism, and his ongoing ambivalence towards the foundational crisis (Atiyah 2003). Max Newman dedicated a section he called "mathematical logic" to Weyl's intuitionistic contemplations, but his piece ends with a functional tone regarding Weyl's turn to intuitionism, as if Weyl's only motivation was to make Brouwer's mathematics accessible to other practitioners (Newman 1957, 323).

On the other hand, Dirk van Dalen claims that Weyl's commitment to intuitionism was not a mere infatuation or transitional stage. Throughout his paper on Weyl's intuitionistic mathematics, van Dalen portrays Weyl as a life-long admirer of Brouwer's foundational approach, who could never quite break with his earlier intuitionistic past (van Dalen 1995). According to van Dalen, the story of Weyl and Brouwer ends with their Zurich meeting, not long before Weyl died, where Weyl remarked sadly: "Brouwer, everything is unsteady again." John Bell tells a similar story of Weyl's lasting fascination with intuitionism, specifically with the notion of the mathematical continuum (Bell 2000).

Erhard Scholz ascribes historical rationality to the moves of Hermann Weyl by taking a "non-systematic" perspective that attempts to elucidate Weyl's changes of mind (Scholz 2000). Scholz describes Weyl's inclination towards intuitionism as rooted in his philosophical considerations, which were deeply influenced by Fichte's approach to the concept of continuum and space. Even though the connection Weyl saw between intuitionism and the established structures of purely infinitesimal geometry may never exist outside his mind, Scholz maintains that illusions have often played a significant role in shaping historical developments, and even if this sense of closeness is later seen as illusory, it must still be acknowledged if we are to understand how Weyl came to play a leading role in promoting intuitionistic analysis (Scholz 2000, 5). Scholz's attempt to take into account Weyl's inner "Besinnung" (translated as "reflection" or "contemplation", see: (Scholz 2004, 14)) as part of his motivation to consider different foundational ideas is a perspective that I wish to expand upon using Fisch's concept of ambivalence.

The retelling of Weyl's story in light of Fisch's notion of ambivalence suggests a view that enhances Bell's, van Dalen's, and Scholz's readings; it adds to their interpretations a deeper philosophical aspect that is currently missing from their accounts and should not be overlooked if one wants to gain a better understating of Weyl's shifting positions. Weyl's ambivalence regarding the foundations of mathematics is a continuous thread that is woven into his writings throughout his life. Thus, his motivation for changing his mind and the roots of his indecision play a significant role within the contours of the whole story.

When Weyl's undecidedness is addressed as ambivalence in a Fischian manner, the rationale behind his constant change of mind unravels. As a practitioner who had been exposed to external criticism (not only in the form of Brouwer's intuitionism, but also through the works of philosophers and mathematicians such as Kant, Fichte, Husserl, Poincare, the French intuitionists, and from 1927 Hilbert again), Weyl found himself engaged in normative self-critical deliberations not only about whether mathematics can be built upon intuitionistic, constructive, or formalistic foundations, but also about what it means to be an intuitionist, a formalist, or a constructivist.

Developments in Brouwer's intuitionism and Hilbert's formalism-finitism also contributed to Weyl's foundational considerations. Neither theory remained stagnant over the years. Brouwer's intuitionistic views changed as he went from addressing the intuition of the continuum as an unanalyzable entity (Brouwer 1907) to introducing new concepts such as spreads and choice sequences in order to show how points on the continuum are identified with specific choice sequences (van Atten 2017; Brouwer 1918). A later development in Brouwer's intuitionism occurred when his proof of fan theorem appeared in 1924, followed by the introduction of continuity theorem dealing with intuitionistic continuous functions (Brouwer 1924, 1927).

Hilbert's views on the foundations of analysis also transformed during the first decades of the twentieth century (Corry 2004), particularly his perspective regarding the logical basis required for rebuilding the foundations of mathematics. Hilbert's quest for consistency proof in the early 1900s led him to abandon Dedekind's logicism (Ferreirós 2009) and turn to Schröder's conception of logic, even though it was not particularly suitable for Hilbert's formalistic purposes (Zach 2019). In 1914, a few years after the publication of Russell and Whitehead's *Principia Mathematica*, Hilbert became intensely involved with this work. He believed that the point of view developed there could lead to laying down the necessary logical foundations for his axiomatic treatment of mathematics (Mancosu 1999). Later on, however, he came to realize that other fundamental problems of axiomatics still remained unsolved. Therefore, he devoted the following years (1917–1921) to developing first-order logic and presented his new foundational approach in 1922 as a response to Brouwer's intuitionism (Hilbert 1922; Zach 2007). Between 1926 and 1928, Hilbert's program was further developed, as he introduced a distinction between real and ideal formulas, followed by his notions of ideal propositions (Hilbert 1926) and real propositions (Hilbert 1927).

In 1925, as the debate gained increasing attention from practitioners inside and outside the field (Hesseling 2003), Weyl published a paper in response to Hilbert's reaction to Brouwer's intuitionism. Despite the paper's conciliatory tone, Weyl criticized Hilbert for his intention to "secure not the *truth*, but the *consistency* of the old analysis" (Weyl 1925, 136). Brouwer's intuitionism offered a very substantial and attractive idea of mathematical truth (Dummett 1975, 1977), but when Brouwer's fan theorem proof was published in 1924, it revealed a problematic aspect in the epistemological standards of Brouwer's intuitionism, threatening the validity and justifiability of the intuitionistic view of mathematical truth (Epple 2000). The realization that Brouwer's intuitionism may not be able to adequately address the notion of mathematical truth as it had promised, alongside the developments in Hilbert's program (Hilbert 1926; Sieg 1999), might have weakened Weyl's confidence in Brouwer's revisionist program and its core concepts, thereby contributing to his feelings of ambivalence.[4]

[4] As can be seen in Sieg's citation of Hilbert from 1923, Hilbert stated that the consistency problem is "no longer a question of proving that a system of infinitely many things is logically possible, but only of recognizing that it is impossible to derive a pair of formulas like A and ~A from the axioms extant in the formulas according to the rules of the logical calculus" (Sieg 1999, 40). Otherwise stated, Hilbert no longer searched for a proof of the logical possibility of an axiomatic system, but rather for a formal proof that an axiomatic system is non-contradicting. Moreover, it shows that

The feeling of ambivalence in Fisch's model accounts for more than merely explaining the persistence of practitioners' hesitant stances or indecisive acts; it primarily motivates for action. Driven by their ambivalence, practitioners are motivated to try and reconcile the stances they are torn between, creating new hybrid solutions. Even though these split accounts often preserve the practitioner's indecision, they serve to ambivalate leading practitioners within the discipline who attempt, in turn, to formulate less jarring accounts.

Within the contours of Weyl's story, I can point to at least two attempts to formulate similar hybridic solutions: one in Weyl's 1918 monograph *Das Kontinuum* and another one three years later in "On the New Foundational Crisis in Mathematics." In *Das Kontinuum* Weyl suggested subjecting the epsilon relation to type restriction in order to avoid the paradoxes of set theory, and to restrict the use of quantifiers in the set-theoretic comprehension scheme in order to avoid the vicious circle of impredicative definitions (Weyl 1918b). The solution he proposed in "On the New Foundational Crisis in Mathematics" was a more radical one, since here Weyl embraced Brouwer's choice sequences, completely forfeiting arithmetically definable sequences and the principle of excluded middle. Both attempts were to some extent misread, and even though *Das Kontinuum* is now regarded as a landmark work, it took nearly sixty years for the scientific community to fully appreciate its insights (van Dalen 2013).

Weyl's frequent shifts in perspective might also be seen as part of a broader historical backdrop of cultural change. Jeremy Gray, in his monograph *Plato's Ghost: The Modernist Transformation of Mathematics*, explores the connection between individual contributions to mathematics and the wider social forces shaping the discipline (Gray 2008). Gray argues that between 1890 and 1930, mathematics underwent a "modernist transformation," a profound cultural shift driven by key intellectual works and the communities that championed them. During this period, groups of influential thinkers had both the resources and the network to promote groundbreaking ideas, sparking a collective movement that reshaped the mathematical landscape. Cultural and political events, particularly the upheaval of the First World War, also played a significant role, influencing the outlook and urgency of mathematicians and their communities (Gray 2008, 406–7).

Even though Gray's analysis mainly focuses on one direction, namely how individuals' works affected the process of transition (and not the other way around), it would be shortsighted to dismiss the way cultural and historical changes affected Weyl's views regarding certain theories, concepts, and ideas. As Dirk van Dalen points out, Weyl himself characterized the tone of his 1921 paper as influenced by "the mood of excited times—the times immediately following the First World War" (Weyl 1956).

This fusion of personal and cultural influence sparks a compelling question: might Weyl's own shifting, ambivalent perspectives have carried the same kind of

---

Hilbert did not regard consistency proofs as indicating anything about mathematical existence, a view that coincides with the intuitionistic approach to consistency.

hybrid, transformative power that George Peacock's *A Treatise on Algebra* wielded in its own era? Fisch suggests that Peacock's work carved out new conceptual spaces by blending old ideas with fresh insights. Inspired by this notion of hybridity, the remainder of this chapter will explore how Weyl's 1921 paper, with its distinctive mix of tradition and innovation, reflects Peacock's pioneering approach in the *Treatise*.

5.5 Rethinking Hybridity: A Juxtaposition of Peacock's and Weyl's Stories

5.5.1 Trading Ideas and Shifting Paradigms

Fisch's theory of normative framework transitions reimagines paradigm shifts in science as two-part processes: in the first stage, established practitioners begin to experience ambivalence. In the second stage, practitioners respond to their ambivalence in a uniquely constructive way: they produce hybrid, "split" accounts of their field, trying to bridge the conflicting norms that now inhabit their thinking (Fisch 2017, 177). Fisch sees these hybrid accounts as powerful tools that reshape fields over time. To illustrate this process, he uses the example of George Peacock's *A Treatise on Algebra*, which he interprets as the carefully reasoned product of Peacock's own struggle to reconcile competing views of algebra.

At the heart of Fisch's theory is the idea that changing one's mind (what he calls "changing one's normative commitments") is rarely a solo journey (Fisch and Benbaji 2011). Altering deeply held beliefs is not something practitioners can achieve on their own; it requires trusted criticism from the outside. Fisch describes this as an "ambivalating" process, where external viewpoints disrupt the practitioner's confidence in their existing framework, initiating a rationally grounded re-evaluation of their core beliefs. But this ambivalating process isn't just personal; it reverberates through the entire field. For an entire discipline to undergo a transformation, two key conditions must be met. First, the field must be exposed to external, unsettling criticism. Second, while the initial push may come from the outside, the real work of transformation must happen from within. It's only when influential practitioners, with both authority and a deep connection to the field, take up this ambivalated state and publish their hybrid insights that genuine change begins to ripple outward.

To explain how a practitioner's commitment to a prevailing framework can be "destabilized," Fisch draws on Peter Galison's compelling metaphor of "scientific trading zones" (Galison 1997). Galison uses the concept of a "trading zone" to shed light on how scientists from different, often conflicting, frameworks can communicate and collaborate. Within these conceptual trading zones, scientists and theorists can exchange tools, techniques, and concepts. They do so by developing shared languages, what Galison calls scientific pidgins and creoles, that mediate their interactions. In this metaphor, theorists and experimenters aren't merely translating concepts but carefully positioning elements of their systems to align with those from other frameworks, navigating across boundaries to create new knowledge (Galison 1997, 803, 816).

Drawing on Galison's framework, Fisch portrays practitioners as traders whose exchanges often go beyond their original intentions: they may encounter unexpected ideas and adopt them if these are seen as filling a gap in their own intellectual toolkit (Fisch 2010). Nevertheless, Fisch points out that practitioners participating in scientific trading zones will often also encounter questions and criticism, something Galison fails to acknowledge.

The mere act of trading, claims Fisch, will not suffice to render someone ambivalent. Traders arrive at the trading zones searching answers to questions arising from within the field of inquiry, in the course of which they often find themselves faced with questions encompassing the difficulties and shortcomings of the framework itself (Fisch 2010, 532). Grappling with the former type of questions, practitioners opt to trade "close to home" as both the questions and the possible solutions are developed from within the prevailing framework. Therefore, not only does such trading never prompt them to criticize their current paradigm, it tends to reinforce their commitment to it. The latter type of questions arises only at the trading zones themselves, where practitioners are liable to be exposed to friendly criticism and questioning of their frameworks. It is only by trading "away from home", when practitioners must articulate fundamental concepts and proofs that are taken for granted in their home field, that a truly destabilizing moment can occur: despite being powerless to convince, external criticism coming from trusted participants encountered in trading zones is still able to provoke its addressees to reconsider their normative commitments.

5.5.2 Hybridity's Many Faces

In Fisch's model, the second stage captures the deeply fractured nature of works born from normative ambivalence. Fisch illustrates this with Peacock, who initially saw algebra as a natural extension of arithmetic, driven by a desire to keep mathematics grounded in meaningful, truth-oriented principles. However, Peacock's exposure to Charles Babbage's entirely formalistic approach to analysis posed a direct challenge to his views. Confronted with this competing vision, Peacock crafted a compromise in the form of "two algebras". Peacock's hybrid solution sought to reconcile Babbage's abstract view with his own reluctance to detach the theory of number from meaningful truths. His response was to distinguish between "arithmetical algebra," which he saw as generalized arithmetic expressed through algebraic symbols, and "symbolic algebra," which aligned with Babbage's view of pure analysis as a self-contained system. In this way, Peacock attempted to balance Babbage's formalism with his own commitment to a meaning-infused mathematical foundation. While he was clearly influenced by Babbage's approach, Peacock was ultimately unable to relinquish his foundational beliefs, resulting in a compromise that maintained the ambivalence at its core.

Weyl's response to Brouwer's intuitionism reveals a similar pattern of hybridization and compromise. Compelled by the need to clarify the elusive concept of the continuum, Weyl was captivated by Brouwer's intuitionistic approach, which seemed to offer a radically new solution to the continuum problem. At first, Weyl was convinced that Brouwer's intuitionism marked a breakthrough, declaring enthusiastically that "a new solution to the continuum problem has been found" (Weyl 1921, 99). However, as much as Weyl admired the promise of Brouwer's ideas, he was not blind to their problematic aspects, such as the restrictive nature of intuitionism, which imposed severe limits on the scope of mathematical practice[5] (van Dalen 1995; Weyl 1921).

Gradually developed out of Brouwer's intuitionism, the theory Weyl presented in his 1921 paper draws a clear distinction between his ideas and Brouwer's theorem, explicitly articulating the concepts he had "borrowed" from Brouwer, alongside those that he is "responsible for" and constitute his new intuitionistic approach (Weyl 1921, 109–10). The problem was that the differences between Brouwer's intuitionism and Weyl's intuitionism were subtle and elusive; a significant level of engagement with both theories was needed in order to tell them apart. Even Hilbert occasionally misattributed some of Weyl's views to Brouwer (van Dalen 1995, 158).

Both Peacock and Weyl, confronted with foundational alternatives that attracted yet conflicted with their beliefs, crafted complex, hybrid solutions that attempted to preserve elements of both their original commitments and the new ideas they found compelling. The structure of Peacock's solution, namely, the attempt to determine the precise relationship between the two algebras, established symbolical algebra as a derivative of arithmetical algebra, much to the contrary of what Peacock originally intended to do. Peacock's hybrid solution consisted of impossible and incoherent welding of his former view (algebra as generalized arithmetic) with a new, modified borrowing from Babbage (algebra as a symbolic formalism). Weyl's solution comprised three fundamental concepts he borrowed from Brouwer (the idea of the developing sequences, the doubt in the principle of excluded middle and the concept of the function continua (Weyl 1921, 109)), that evolved into a new theory of the continuum as an entity that cannot be divided into separate fragments (as opposed to Cantor's view of the continuum). Weyl's theory was not impossible nor hybrid in the sense that Peacock's solution was; even though Weyl did not fully endorse

[5] A significant restriction of Brouwer's intuitionism on mathematics was its complete rebuttal of the concept of sets of numbers and sets of sets. As Weyl himself stated, "there is no place in our analysis for a general set theory, as little as there is room for general statements about functions. The new conception, as one can see, brings with it far-reaching restrictions [...]" (Weyl 1921, 109). In a footnote, Weyl cautiously explained that general statements about sets and functions are not impossible, but they are not based on an intuitive foundation (unlike the generality of arithmetic and analysis, that are based in each case on a distinct intuitive foundation and as a result contain independent, intuitive content). Even though it is possible to form such general propositions into a field called "set theory," it cannot constitute the foundation of mathematics from an intuitionistic point of view (Weyl 1921, 117).

Brouwer, his newly proposed theory aligned with much of Brouwer's ideas and was deprived of any attempt to combine them with Weyl's former views. The new analysis Weyl presented in his 1921 paper was utterly isolated from any other mathematical theory (besides Brouwer's intuitionism).

The retelling of Hermann Weyl's story in light of Fisch's model reveals that the cases of Weyl and Peacock have two things in common: the first is a feeling of ambivalence, which led them to form new and creative solutions. The second characteristic they share is a determination to find solutions for what they perceive as central and notably essential problems in mathematics. Just as Peacock steadfastly pursued a new solution that would bridge the two jarring conceptions of algebra, Weyl was ardent to find a new analysis that "conforms to the intuitive character of the continuum" (Weyl 1921, 117).

Such a reading also brings to the surface the differences between the two stories. As opposed to Peacock's *Treatise*, Weyl's 1921 paper was not an impossible amalgam of two theories about the foundations of mathematics, nor did it prompt other practitioners to take a more decisive stand. Peacock's work had succeeded in prompting four major figures to respond (Augustus De Morgan, William Rowan Hamilton, Duncan Farquharson Gregory, and William Whewell), and even though they all misread the *Treatise* in light of their own perspectives, all four were rendered ambivalent and felt compelled to respond. That was not the case in Weyl's story. Not long after the initial resonance to Weyl's 1921 paper had started to dissolve, practitioners of voice and standing have found themselves reluctant to engage with intuitionism as introduced by Weyl, thereby shifting the direction of the debate towards other semi-intuitionistic streams such as Heyting's formalization of intuitionism (Hesseling 2003).

5.6 Concluding Remarks: Weyl's Lifelong Dance with Ambivalence

Understanding Hermann Weyl's enduring ambivalence about the foundations of mathematics requires stepping beyond traditional historical accounts and into the realm of philosophical interpretation, particularly Menachem Fisch's notion of normative ambivalence. This chapter presented an alternative reading of Weyl's turn to intuitionism, suggesting that a philosophical perspective of normative framework transitions is fundamentally missing from the current historical accounts of Weyl's story. Specifically, it claimed that the key to understanding Weyl's undecidedness lies in Fisch's notion of ambivalence.

Fisch maintains that we are capable of becoming ambivalent, but not to ambivalating ourselves towards our norms. Only a trusted criticism coming from without can change our whole point of view and induce a rational, justified transformation of our normative framework. The process of becoming and being ambivalent has no time restrictions and can last throughout a practitioner's professional career.

Moreover, there is no guarantee that ambivalence will eventually lead to a clear decisive position; a practitioner might remain ambivalent during his whole life, and Weyl is a good case in point.

Unlike other practitioners who may have experienced the same difficulties regarding the foundational crisis, Weyl expressed his prolonged search for a solution in his writings, explicitly articulating the problematic aspects of every mathematical theory he had considered. The continued tone of ambivalence present in his papers is not a common phenomenon within the mathematical landscape, and it was often attributed to his philosophical inclinations that also changed over the years. The philosophical perspective presented in this chapter suggests ways of providing a different analysis of Weyl's continuous undecidedness, taking into consideration the intrasubjective process of changing one's normative framework. Weyl's so-called shifting positions should be regarded only as symptoms of a much deeper, convoluted intrapersonal process of self-deliberation, in his attempt to find a solid ground on which mathematics can be built.

Seen through Fisch's theory, Weyl's "shifting positions" signal his attempt to navigate the vast landscape of mathematics while remaining open to its uncertainties, rather than forcing resolution where he found none. Weyl's legacy, therefore, is not just in the theories he advanced but in the way he exemplified a rare intellectual openness, allowing his work to reflect his continuous, often uncomfortable questioning. By interpreting Weyl's journey through this philosophical lens, we see a mathematician not just wrestling with numbers and proofs but with the very nature of knowledge, certainty, and the possibility of truth in mathematics.

References

Atiyah, Michael. 2003. "Hermann Weyl" biographical. *Memoirs* 82:320–335.

Beisswanger, Peter. 1965. The phases in Hermann Weyl's assessment of mathematics. *Mathematic-Physical Semester Reports* 12:132–156.

Bell, John. 2000. Hermann Weyl on intuition and the continuum. *Philosophia Mathematica* 3(8):259–273.

Bell, John. 2004. Hermann Weyl's later philosophical views: His divergence from Husserl. In *Husserl and the sciences: Selected perspectives*, ed. Richard Feist, 173–186. Ottawa, ON: University of Ottawa Press.

Bell, John, and Herbert Korté. 2015. Hermann Weyl. In *The Stanford Encyclopedia of philosophy*. Stanford University.

Brouwer, Luitzen Egbertus Jan. 1907. *On the foundations of mathematics*. Ph.D. Dissertation, University of Amsterdam.

Brouwer, Luitzen Egbertus Jan. 1912. Intuitionism and formalism. In *L.E.J Brouwer collected works I—Philosophy and foundations of mathematics*, ed. Arend Heyting, 123–138. Amsterdam: North-Holland.

Brouwer, Luitzen Egbertus Jan. 1918. Founding set theory independently of the principle of the excluded middle. *KNAW Verhandelingen* 5:1–43.

Brouwer, Luitzen Egbertus Jan. 1924. Über Die Bedeutung Des Satzes Vom Ausgeschlossenen Dritten in Der Mathematik, Insbesondere in Der Funktionentheorie. In *From Frege to Gödel. A sourcebook in mathematical logic, 1879–1931*, ed. Jean van Heijenoort, 335–341. Cambridge, MA: Harvard University Press.

Brouwer, Luitzen Egbertus Jan. 1927. Über Definitionsbereiche von Funktionen. In *From Frege to Gödel. A sourcebook in mathematical logic, 1879–1931*, ed. Jean van Heijenoort, 457–463. Cambridge, MA: Harvard University Press.

Corry, Leo. 2004. *David Hilbert and the axiomatization of physics (1898–1918). From Grundlagen Der Geometrie to Grundlagen Der Physik*. London: Springer.

Dummett, Michael. 1975. The philosophical basis of intuitionistic logic. In *Truth and other enigmas*, 215–247. Cambridge, MA: Harvard University Press.

Dummett, Michael. 1977. *Elements of intuitionism*. Oxford: Clarendon Press.

Epple, Moritz. 2000. Did Brouwer's intuitionistic analysis satisfy its own epistemological standards? In *Proof theory: History and philosophical significance*, ed. V. F. Hendricks et al., 153–178. Dordrecht: Kluwer.

Feferman, Solomon. 1988. The development of programs for the foundations of mathematics in the first third of the 20th century. https://pdfs.semanticscholar.org/5a3a/112929a5a2a1f1fec4447cc584628918c0fb.pdf

Feferman, Solomon. 1998. *In the light of logic*. New York: Oxford University Press.

Feferman, Solomon. 2000. The significance of Weyl's Das Kontinuum. In *Proof theory: History and philosophical significance*, ed. V. F. Hendricks et al., 179–194. Dordrecht: Kluwer.

Feist, Richard. 2002. Weyl's appropriation of Husserl's and Poincaré's thought. *Synthese* 132 (3): 273–301.

Ferreirós, Jose. 2009. Hilbert, Logicism, and mathematical existence. *Synthese* 170 (1): 33–70.

Fisch, Menachem. 2010. Towards a history and philosophy of scientific agency. *Monist* 93 (4): 518–544.

Fisch, Menachem. 2017. *Creatively undecided: Toward a history and philosophy of scientific agency*. Chicago: The University of Chicago Press.

Fisch, Menachem, and Yitzhak Benbaji. 2011. *The view from within: Normativity and the limits of self-criticism*. Notre Dame: University of Notre Dame Press.

Galison, Peter. 1997. *Image and logic: A material culture of microphysics*. Chicago and London: The University of Chicago Press.

Gray, Jeremy. 2008. *Plato's ghost: The modernist transformation of mathematics*. Princeton: Princeton University Press.

Hesseling, Dennis E. 2003. *Gnomes in the fog: The reception of Brouwer's intuitionism in the 1920s*. Basel, Boston, Berlin: Birkhäuser Verlag.

Hilbert, David. 1922. Neubegründung Der Mathematik: Erste Mitteilung. In *From Brouwer to Hilbert: The debate on the foundations of mathematics in the 1920s*, ed. Paolo Mancosu, 149–188. Oxford: Oxford University Press.

Hilbert, David. 1926. Über Das Unendliche. In *From Frege to Gödel. A sourcebook in mathematical logic, 1879–1931*, ed. Jean van Heijenoort, 367–392. Cambridge, MA: Harvard University Press.

Hilbert, David. 1927. The foundations of mathematics. In *From Frege to Gödel: A source book in mathematical logic, 1879–1931*, ed. Jean van Heijenoort, 464–479. Cambridge, MA: Harvard University Press.

Mancosu, Paolo. 1999. Between Russell and Hilbert: Behmann on the foundations of mathematics. *The Bulletin of Symbolic Logic* 5 (3): 303–330.

Mancosu, Paolo, and Thomas Ryckman. 2002. Mathematics and phenomenology: The correspondence between O. Becker and H. Weyl. *Philosophia Mathematica* 10 (2): 130–202.

Mannoury, Gerrit. 1909. *Methodologisches Und Philosophisches Zur ElementarMathematik*. Haarlem: P. Visser Azn.

Moore, Gregory H. 1984. *Zermelo's axiom of choice*. Berlin: Springer.

Newman, Max. 1957. Hermann Weyl, 1885-1955. *Biographical Memoirs of Fellows of the Royal Society* 3:305–328.

Poincaré, Henri. 1963. *Mathematics and science: Last essays*. New York: Dover Press.

Rosello, Joan. 2012. *From foundations to philosophy of mathematics: An historical account of their development in the XX century and beyond.* Cambridge, MA: Cambridge Scholars Publishing.

Scholz, Erhard. 2000. Herman Weyl on the concept of continuum. In *Proof theory: History and philosophical significance*, ed. V. F. Hendricks et al., 195–220. Dordrecht: Kluwer.

Scholz, Erhard. 2004. Philosophy as a cultural resource and medium of reflection for Hermann Weyl. https://arxiv.org/abs/math/0409596

Sieg, Wilfried. 1999. Hilbert's programs: 1917-1922. *The Bulletin of Symbolic Logic* 5:1–44.

Sieroka, Norman. 2007. Weyl's 'Agens theory' of matter and the Zurich Fichte. *Studies in History and Philosophy of Science Part A* 38 (1): 84–107.

Sieroka, Norman. 2009. Husserlian and Fichtean leanings: Weyl on Logicism, intuitionism, and formalism. *Philosophia Scientiae* 13 (2): 85–96.

Sieroka, Norman. 2019. Neighbourhoods and Intersubjectivity: Analogies between Weyl's analyses of the continuum and transcendental-phenomenological theories of subjectivity. In *Weyl and the problem of space: From science to philosophy*, ed. Julien Bernard and Carlos Lobo, 98–122. Cham, Switzerland: Springer.

van Atten, Mark. 2017. The development of intuitionistic logic. In *The Stanford Encyclopedia of Philosophy (Winter 2017 Edition)*. Stanford University.

van Dalen, Dirk. 1995. Hermann Weyl's intuitionistic mathematics. *The Bulletin of Symbolic Logic* 1 (2): 145–169.

van Dalen, Dirk. 1999. *Mystic, geometer, and intuitionist: The life of L.E.J. Brouwer. Volume I: The dawning revolution*. Oxford: Clarendon Press.

van Dalen, Dirk. 2013. *L.E.J. Brouwer—Topologist, intuitionist, philosopher: How mathematics is rooted in life*. London: Springer.

van Stigt, Walter. 1990. *Studies in the history and philosophy of mathematics: Brouwer's intuitionism*. Amsterdam: North-Holland.

Weyl, Hermann. 1910. "Über Die Definitionen Der Mathematischen Grundbegriffe" Mathematisch-naturwissenschaftliche. *Blätter* 7:93–95.

Weyl, Hermann. 1913. *Die Idee Der Riemannschen Fläche*. Leipzig: B. G. Teubner.

Weyl, Hermann. 1918a. *Space-time-matter*. New York: Dover.

Weyl, Hermann. 1918b. *The continuum: A critical examination of the foundation of analysis*. Thomas Jefferson University Press.

Weyl, Hermann. 1921. On the new foundational crisis in mathematics. In *From Brouwer to Hilbert: The debate on the foundations of mathematics in the 1920s*, ed. Paolo Mancosu, 86–118. Oxford: Oxford University Press.

Weyl, Hermann. 1924. Das Gruppentheoretische Fundament Der Tensorrechnung. *Mathematisch-physikalische Klasse* 1924:218–224.

Weyl, Hermann. 1925. The current epistemological situation in mathematics. In *From Brouwer to Hilbert: The debate on the foundations of mathematics in the 1920s*, ed. Paolo Mancosu, 123–142. Oxford: Oxford University Press.

Weyl, Hermann. 1926a. Theorie Der Darstellung Kontinuierlicher Halbeinfacher Gruppen Durch Lineare Transformationen II. *Mathematische Zeitschrift* 24:328–376.

Weyl, Hermann. 1926b. Theorie Der Darstellung Kontinuierlicher Halbeinfacher Gruppen Durch Lineare Transformationen III. *Mathematische Zeitschrift* 24:377–395.

Weyl, Hermann. 1926c. Theorie Der Darstellung Kontinuierlicher Halbeinfacher Gruppen Durch Lineare Transformationen (Nachtrag). *Mathematische Zeitschrift* 24:789–791.

Weyl, Hermann. 1928. *Gruppentheorie Und Quantenmechanik*. S. Hirzel.

Weyl, Hermann. 1931. *Die Stufen Des Unendlichen*. Jena: Gustav Fischer.

Weyl, Hermann. 1934. *Mind and nature*. Philadelphia, PA: University of Pennsylvania Press.

Weyl, Hermann. 1939. *The classical groups, their invariants and representations*. Princeton: Princeton University Press.

Weyl, Hermann. 1940. The mathematical way of thinking. *Science* 92 (2394): 437–446.

Weyl, Hermann. 1944. *Meromorphic functions and analytic curves*. Princeton: Princeton University Press.

Weyl, Hermann. 1946. Mathematics and logic. *American Mathematical Monthly* 53:2–13.
Weyl, Hermann. 1949. *Philosophy of mathematics and natural science*. Princeton: Princeton University Press.
Weyl, Hermann. 1950. *Group theory and quantum mechanics*. Dover.
Weyl, Hermann. 1951. A half-century of mathematics. *The American Mathematical Monthly* 58 (8): 523–553.
Weyl, Hermann. 1952. *Symmetry*. Princeton, NJ: Princeton University Press.
Weyl, Hermann. 1953. Über Den Symbolismus Der Mathematik Und Mathematischen Physik. *Stadium General* 6:219–228.
Weyl, Hermann. 1954a. *Algebraic theory of numbers*. Princeton, NJ: Princeton University Press.
Weyl, Hermann. 1954b. Axiomatic versus constructive procedures in mathematics. *Mathematical Intelligencer* 7 (4): 10–38.
Weyl, Hermann. 1955. Insight and reflection. In *The Spirit and uses of the mathematical sciences*, ed. T. L. Saaty and F. J. Weyl, 281–301. New York: McGraw-Hill.
Weyl, Hermann. 1956. *Selecta Hermann Weyl*. Basel: Birkhäuser.
Zach, Richard. 2007. Hilbert's program then and now. In *Philosophy of logic*, ed. Dale Jacquette, 411–447. Amsterdam: North-Holland.
Zach, Richard. 2019. Hilbert's program. In *The Stanford Encyclopedia of philosophy*. Stanford University. https://plato.stanford.edu/archives/sum2019/entries/hilbert-program/.

Chapter 6
Transcending Communal Boundaries: Brouwer's Intuitionism in Philosophical Discourse

Abstract This chapter explores the divergent receptions of Brouwer's intuitionism within the fields of mathematics and philosophy. While Brouwer's ideas largely faded from mainstream mathematical practice by the mid-twentieth century, they continue to hold a significant place in philosophical discussions on the foundations of mathematics. Through an analysis of influential works in the philosophy of mathematics, the chapter reveals how philosophers have upheld intuitionism as a profound perspective on mathematical truth and existence, even as mathematicians moved on to frameworks that better served practical needs. This chapter considers the historical and conceptual reasons behind this disparity, suggesting that intuitionism's philosophical depth and challenge to conventional ideas about knowledge and reality have preserved its relevance for philosophers. By examining how intuitionism traversed disciplinary boundaries, this chapter highlights philosophy's unique role in preserving and advancing ideas that may be sidelined within their original scientific domains, ultimately underscoring the distinct standards, norms, and purposes that shape how each community engages with new theories.

Keywords Logicism, formalism, and intuitionism · Scientific communities · Philosophy of mathematics · Mainstream mathematics · Marginalization of theories · Intuitionism's reception

As far as mathematics is concerned, Brouwer's intuitionistic program did not win the larger mathematical battle (Posy 2005, 319). Although intuitionism as a whole was never completely abandoned, the mathematical doctrines that have taken on independent lives, such as those of Heyting and his students, were developed in ways that diverge significantly from Brouwer's original endeavor. Brouwer's intuitionism ceased to be discussed sometime after the 1930s, as Heyting stirred the foundational debate towards his own intuitionistic approach.

However, upon browsing the pages of eminent volumes dedicated to the philosophy of mathematics, a different picture emerges. Dating back to Benacerraf and Putnam's monumental volume from 1964 (Putnam and Benacerraf 1964), and all the way through more recent publications, such as Stewart Shapiro's *Thinking about*

© The Author(s), under exclusive license to Springer Nature Switzerland AG 2025

K. Kish Bar-On, *Intuitionism Resocialized*, Studies in History and Philosophy of Science 66, https://doi.org/10.1007/978-3-032-05839-3_6

Mathematics: The Philosophy of Mathematics (Shapiro 2000) and Alexander George and Daniel Velleman's *Philosophies of mathematics* (George and Velleman 2002), Brouwer's intuitionism is treated as if it was never defeated in the mathematical battlefield. That is not to claim that these works ignore the mathematical community's lack of acceptance of Brouwer's intuitionism; they certainly acknowledge that. Nevertheless, each of these works, as well as several other collected readings, address Brouwer's intuitionism as one of the three or four central streams in the philosophy of mathematics. An outsider encountering one of these volumes for the first time might think that in the wider philosophical context, Brouwer's intuitionism is very much alive. Would he be wrong?

The purpose of the current chapter is to shift the focus from the reaction of the mathematical community and individual mathematicians to philosophers and the philosophical community. Throughout the chapter, I examine the position intuitionism came to occupy in several philosophical volumes dedicated to the philosophy of mathematics, as well as intuitionistic works of several prominent philosophers of mathematics. It will show that when changing the perspective from a mathematical one to a philosophical one, the impression left by Brouwer's intuitionism turned out to be so profound that philosophers overlooked, at least to some extent, the fact that in mathematics, intuitionism eventually did not prevail.

6.1 Intuitionism's Place in the Philosophical Landscape

In *The Philosophy of Mathematics: An Introductory Essay*, published in 1962, Stephan Körner presents a comprehensive overview of the main currents in mathematical philosophy (Körner 1962). Among the book's eight chapters, Chapters VI and VII focus on mathematics as the activity of intuitive constructions, directly engaging with intuitionism. This choice is notable, as by 1962, Brouwer's intuitionism had largely faded from the forefront of mathematical practice. In the book's preface, Körner acknowledges the diverse motivations behind each chapter, explaining that Chapters I, II, IV, and VI are devoted to exposition of various views which are either "historically important" or "widely held at the present time." Since intuitionistic ideas were not "widely held" in the mathematical community by the 1960s, it seems clear that, in Körner's view, intuitionism deserves attention due to its historical importance. This raises an interesting question: why would a philosopher like Körner dedicate significant space to a perspective that had largely fallen out of favor within mathematics itself?

The answer, Körner suggests, lies in his distinction between mathematics and the philosophy of mathematics. The central aim of Körner's work, so he claims, was to act as a reminder that "philosophy of mathematics is neither mathematics nor mere popularization of mathematics" (Körner 1962, 186). According to Körner, "the philosophy of mathematics does not add to the number of mathematical theorems and theories. It is not mathematics. It is reflection upon mathematics, giving rise to its own particular questions and answers" (Körner 1962, 9). The differentiation between mathematics and philosophy of mathematics that Körner points to explains, at least to some extent, why intuitionism continued to be discussed in philosophy long after its renunciation by mathematicians: in philosophy of mathematics, intuitionism is another approach used to contemplate the problems of mathematics as a discipline without adding or changing anything in the body of mathematical knowledge. Hence, intuitionism plays an entirely different role in philosophy than it does in mathematics, where it aimed to change, discard, and add new theorems.

Körner was not alone in treating intuitionism as one of the essential currents in the philosophy of mathematics. Just a couple of years after his book, in 1964, Hilary Putnam and Paul Benacerraf compiled a volume titled *Philosophy of Mathematics: Selected Readings* that would go on to become a cornerstone in the field of mathematical philosophy. In the first part of this collection, intuitionism stands as an equal counterpart to the other schools, given serious attention as a philosophical approach that engages with mathematics on a fundamental level. By the time the second edition was published in 1984, Putnam and Benacerraf not only kept the focus on intuitionism but strengthened it, adding Michael Dummett's paper on the philosophical basis of intuitionistic logic. This addition was more than a simple update; it showed a commitment to maintaining intuitionism's position as a key approach in philosophical discussions of mathematics.

In their preface, Benacerraf and Putnam are explicit about the intentions behind their editorial choices. They clarify that they did not attempt to compile a collection that could be said to cover the entire field of the philosophy of mathematics in a comprehensive way. Instead, their goal was to gather selections they found "interesting in their own right" and that, when read together, offered meaningful comparisons (Putnam and Benacerraf 1984, 1). They also emphasize that they didn't try to present a unified perspective across the anthology. In their words, the volume speaks from many, different, and often incompatible viewpoints (Putnam and Benacerraf 1984, 5). This choice was intentional for two main reasons: first, they wanted to raise useful and interesting questions rather than providing definitive answers. Second, they acknowledged that finding a single, unified perspective would be difficult, if not impossible, given the diversity and depth of the issues and viewpoints.

Much like Körner, Putnam and Benacerraf saw the philosophy of mathematics as less a quest for answers than a platform for asking big, sometimes uncomfortable questions. To them, intuitionism stood side-by-side with logicism and formalism, not as a rival theory to be proven or disproven, but as a lens through which to examine the deeper issues that lie at mathematics' foundation. They referred to intuitionism, logicism, and formalism as "epistemologies of mathematics," each one offering a unique view of what it means for mathematical knowledge to be valid.

The opening chapters of their collection take readers back to a historic moment in mathematical thought: the 1930 Königsberg conference, where three giants of the field—Arend Heyting, Rudolf Carnap, and John von Neumann—each presented a lecture that captured his view on constructivity and the nature of mathematical existence.

Carnap defended logicism and argued that all mathematical notions could be reduced to logical notions by explicit definition (Carnap 1931). Thus, implicit definitions (that occur in formalism) do not occur in logicism or intuitionism, which makes both better candidates for providing solid foundations to mathematics. The thing that makes logicism more suitable to be the basis of mathematics than intuitionism, so claims Carnap, is that logicism would accept second-order predicate logic whereas intuitionism would not (Carnap 1931, 50–2).

John von Neumann presented a compelling case for formalism, framing it as a robust, internally consistent system (von Neumann 1931). He argued that even if classical mathematics were eventually found to lack real-world "truth," it remained valuable due to its closed, rule-bound structure. In von Neumann's view, formalism consisted of step-by-step procedures, where mathematicians construct valid expressions by combining symbols in accordance with well-defined rules (von Neumann 1931, 61–62). This process was both "finitary" and constructively sound, creating a self-contained mathematical universe based on its internal logic rather than on any external reality.

However, Gödel's short lecture at the end of the conference turned things upside down. While von Neumann maintained the firm belief that logic should play a key role in mathematical praxis, the presentation of Gödel's results indicated that it is impossible to obtain complete proof of the logical coherence of mathematics. It was a death blow to Hilbert's program and to von Neumann's support, which ultimately made him abandon his work on mathematical logic (Israel and Gasca 2009, 29–48).

Heyting defended intuitionism by providing an overview of its core principles: the concept of unity, choice sequences, spread, species, the undivided continuum, and the problematic law of excluded middle. As for mathematical existence, Heyting argued that mathematical objects depend inherently on human thought, and their existence is secured only to the extent that they can be constructed or defined through mental activity (Heyting 1931, 53). His views on constructivity in mathematics are reflected in his closing remark: "intuitionism is the only possible way to construct mathematics" (Heyting 1931, 61).

The setting of these three positions as the leading positions at the Königsberg conference, and later in the proceedings in *Erkenntnis* journal and Putnam and Benacerraf's work, shaped the discussion from 1931 onwards. After 1931, the image most people had of the foundational debate in the 1920s and 1930s was that it engaged three major schools of thought: intuitionism, formalism, and logicism, and that those positions were stable and inflexible. However, until the Königsberg conference, the distinction between the different positions was not that clear (Hesseling 2003; Dawson 1984). Moreover, as Heyting claimed in his paper about the status of mathematics, written thirty years after the Königsberg conference, "none of the conceptions of mathematics is today as clear-cut as it was in 1930" (Heyting 1966, 195). According to Heyting, the difference between the situation in 1930 and 1966 is that "the spirit of peaceful cooperation has gained the victory over that of ruthless contest", in the sense that the distinguishing lines between prominent mathematical streams became much less salient over the years, and the trajectory of the debate moved towards a synthesis between them.

Here arises yet another difference between the way philosophers and mathematicians viewed the evolution of the debate. Heyting described the situation from mathematicians' point of view; but as the focus shifted to the philosophical community, even Heyting acknowledged philosophers' tendency to demarcate the mathematical schools, stating that the philosophical significance of foundational research in mathematics lies, at least in part, in its ability to distinguish between the formal, intuitive, logical, and Platonist components of classical mathematics, and to precisely define both their scope and their limitations (Heyting 1966, 194).

The three positions introduced at the Königsberg conference were canonized by the philosophical community, at least to some extent. Körner, Putnam, and Benacerraf are not the only scholars who gave the three currents a great share of attention; recent publications that engage different philosophies of mathematics follow the same path. One such example is Alexander George and Daniel Velleman's book *Philosophies of Mathematics,* in which the authors devote two out of eight chapters to intuitionism and intuitionistic mathematics, a chapter to logicism, one to finitism, and another to Gödel's incompleteness theorems (George and Velleman 2002).

George and Velleman address the foundational debate as a rare pause in which mathematicians stopped to think and discuss what they are doing and describe it as one of the most inspiring "intellectual adventures" of the twentieth century (George and Velleman 2002, vi). As the debate gradually faded, there was a relative calm about foundational issues, which is "now again the norm within the mathematical community" (George and Velleman 2002, vi). According to George and Velleman's view, the fact that the mathematical community no longer discusses intuitionistic ideas is by no means an indication that the philosophical community should cease to do so as well. On the contrary—they argue that "while the central philosophical issues that divided the protagonists remain, we have been left a rich treasure of ideas, philosophical and mathematics, with which to continue the inquiry" (George and Velleman 2002, vi).

In their book, George and Velleman focus on three foundational programs in mathematics: intuitionism, logicism, and finitism. They argue that you can't fully appreciate any one of these perspectives without considering it in relation to the others, and that these foundational approaches are closely linked to some of the most influential ideas in the philosophy of mathematics that emerged later. Each represents a bold attempt to grapple with the nature of mathematics, to define and perhaps even contain its vastness. Understanding contemporary philosophical discussions, they suggest, requires a look back at these historical efforts, as they provide the essential backdrop that shapes and informs modern thought in the field (George and Velleman 2002, vii).

Another comprehensive collection of essays covering almost every position in the philosophy of mathematics is Stewart Shapiro's *The Oxford handbook of Philosophy of Mathematics and Logic* (Shapiro 2005). Intuitionism is featured in this volume as part of "the big three", namely, the three prominent philosophical positions that dominated the foundational debate in the early twentieth century and continue to define key points in contemporary literature (Shapiro 2000, viii). The other two are logicism and formalism, and even though it discusses a wide range of philosophies of mathematics, the big three comprise the lion's share of the book.

The exploration of intuitionism in Shapiro's volume begins with Carl Posy's chapter, "*Intuitionism and Philosophy*" (Posy 2005). Posy opens his discussion by emphasizing that intuitionism's philosophical aspects cannot be separated from its underlying technical framework. Following Posy's contribution, David Charles McCarty's chapter, "*Intuitionism in Mathematics*", reinforces this view (McCarty 2005). McCarty argues that, while philosophical reflections on intuitionism can offer valuable insights, they should not detract from intuitionism's primary focus: constructing a mathematical framework that stands in opposition to conventional mathematics, based on its own distinct mathematical principles. The mathematical aspect of intuitionism is not only inseparable but also fundamental to understanding the philosophical one, at least in the eyes of Posy and McCarty. The fact that this volume was published almost a century after the first appearance of Brouwer's intuitionism, during which it was marginalized to the extent of being discussed mainly in very small circles of Brouwer's and Heyting's students, seems not to effect how philosophers engage it.

In 2004, two international conferences were held at Uppsala University, where philosophers and mathematicians met to discuss logicism, intuitionism, formalism, and constructive mathematics. The product of these conferences, whose primary purpose was to assess the relevance of those foundational programs to contemporary philosophy of mathematics, amounted to a collection of papers titled *Logicism, Intuitionism and Formalism: What has become of them?* (Lindström et al. 2009). The purpose of the volume, claim its four editors (representing both philosophy and mathematics departments), is to examine to "what extent do the classical programmes of logicism, intuitionism and formalism represent options that are still alive today" (Lindström et al. 2009, 1). The variety of topics and contributors concerned with intuitionism and constructive mathematics throughout the volume indicate that, at the very least, intuitionistic ideas and their developments remain relevant to both philosophers and mathematicians today.

James Robert Brown has also dedicated a book to introduce contemporary philosophies of mathematics, in which one chapter is devoted to Brouwer's intuitionism (Brown 2008). The chapter, titled "Constructive Approaches", presents Brouwer as the father of modern constructive mathematics (Brown 2008, 120). Contemporary constructive approaches, Brown argues, have a "greater debt to his intuitionism than to anything else" (Brown 2008, 120). However, as Brown's book is more concerned with mathematical proofs and his concept of "mathematical image", Brouwer's intuitionism receives considerably less attention than it does in any of the works mentioned above. As a matter of fact, there are only three pages in the whole book that discuss Brouwer's intuitionism per se.

In surveying the landscape of modern philosophy of mathematics, it becomes clear that some scholars intentionally sidestep intuitionism altogether. For instance, *New Waves in Philosophy of Mathematics*, edited by Otávio Bueno and Øystein Linnebo, serves as a telling example of this selective focus (Bueno and Linnebo 2009). In their introduction, Bueno and Linnebo characterize intuitionism as a relic of the "recent past," marking a period when discussions around the three classical philosophies—intuitionism, logicism, and formalism—were central to the field

(Bueno and Linnebo 2009, 2). Now, however, they see the philosophical conversation moving beyond these foundational approaches, relegating intuitionism to just a few brief mentions throughout the volume. In a contribution to the same book, Mary Leng remarks that, a century after the foundational crisis, philosophers are "generally much less worried about the foundations of mathematics," reflecting a broader shift away from foundationalist approaches across philosophical disciplines (Leng 2009, 117).

Similarly, Mark Colyvan's book on the philosophies of mathematics also keeps intuitionism at arm's length (Colyvan 2012). Although he acknowledges that philosophy can illuminate the nature of mathematics, he chooses to minimize discussions on intuitionism. His treatment of intuitionism appears in a section titled "Skipping Through the Big Isms," signaling his intention to leap beyond the old debates of formalism, intuitionism, and logicism (Colyvan 2012, 3). According to Colyvan, after decades of engaging with these foundational debates, philosophers and mathematicians alike grew weary of them, and foundational issues in mathematics, while still present, have lost much of their original urgency and driving purpose.

The fact that some philosophers address intuitionism as an inseparable part of the philosophy of mathematics while others intentionally refrain from engaging with it only accentuates the difference between how the philosophical and the mathematical communities reacted to Brouwer's intuitionistic program. Unlike the mathematical community, in which the discussion of intuitionism gradually ground to a halt, in the philosophical community, even those who deliberately abstain from discussing intuitionism note its existence and feel obliged to explain why they see fit to avoid the subject. Once intuitionism had established its place as one of the "big three" or the "big isms", disregarding it in relevant philosophical discussion demanded an explanation regardless of its demise. In mathematics, the situation is reversed: it is the *inclusion* of by now defunct intuitionistic ideas in mainstream textbooks that is considered a deviation from the norm, that needs to be accounted for.

Beyond its recognition in philosophical overviews, Brouwer's intuitionism also served as fertile ground for prominent philosophers (like Michael Dummett) to develop unique theories rooted in intuitionistic principles. Unlike the mathematical trajectory of intuitionism, which largely continued through Brouwer's direct students and their intellectual descendants, intuitionism's philosophical evolution branched out. These new developments weren't closely tied to Brouwer's inner circle but emerged from philosophers in various professional backgrounds. The following section will explore their contributions and examine the broader impact they've had on the field.

6.2 The Philosophers Who Kept Intuitionism Alive

One of the most important intuitionistic developments in philosophy, and perhaps the most eminent, is Michael Dummett's work on intuitionistic logic and its underlying philosophical basis (Dummett 1975; Iemhoff 2020). Dummett project revolved around the question of whether it was possible to create a philosophical foundation to logic that would replace classical reasoning with intuitionistic reasoning. Following Wittgenstein, Dummett's theory focused on the concept of meaning, specifically on the idea that the meaning of a sentence is determined by how the sentence is used (Dummett 1978, 215–16). In particular, it connected meaning with knowledge and understanding: an individual who knows the meaning of a statement understands it, and vice versa (Dummett 1991).

Dummett's work, however, was not a straightforward continuation of Brouwer's ideas. While Brouwer regarded intuitionistic mathematics as a pre-linguistic, language-less activity, Dummett took a different path. He rooted his philosophy in the connection between meaning, knowledge, and understanding, a framework that assumes the central role of language in making sense of mathematical concepts. For Dummett, classical logic fell short because it relied on a notion of truth detached from any requirement of verification or understanding by the user, whereas intuitionistic logic, with its emphasis on constructive proof and verifiability, aligned more closely with an acceptable meaning theory.

Dummett's paper, "The Philosophical Basis of Intuitionistic Logic", first appeared in the proceedings of the 1973 logic colloquium and was later republished in the second edition of Benacerraf and Putnam's influential anthology on the philosophy of mathematics in 1984 (Putnam and Benacerraf 1984). In their preface, Benacerraf and Putnam explained that the revisions they made between the first and second editions mirrored their sense of how the discipline had evolved over the previous two decades, with Dummett's contributions marking a significant turning point (Putnam and Benacerraf 1984, vii). Indeed, his work ignited a vigorous debate that would extend through the mid-1980s and into the 1990s, provoking responses from mathematicians and philosophers alike.

In their book *Constructivism in Mathematics*, mathematicians Dirk van Dalen and Anne Troelstra recommend Dummett's *Elements of Intuitionism* (Dummett 1977) as an introduction to intuitionism that includes only a small amount of mathematics, but offers substantial content in logic, metamathematics, and philosophical analysis (Troelstra and van Dalen 1988, 30). Nonetheless, later in the book, they criticize Dummett's intuitionistic approach as too narrow since it focuses only on logic and arithmetic, thus not taking into account other mathematical fields.

Alexander George and Alan Weir embarked on a detailed, back-and-forth exchange over the finer points of Dummett's theory, with each tackling different aspects of his approach to meaning and mathematical cognition (George 1987, 1988, 1993; Weir 1986, 1988). Both critics found themselves skeptical of Dummett's ideas, though their objections diverged significantly. Weir argued against Dummett's intuitionistic stance, suggesting that, even if one fully accepted Dummett's theory of meaning, there was still no compelling reason to discard classical logic when dealing with statements that have a clear and determinate meaning (Weir 1986, 466). George, however, found Weir's objections lacking, dismissing them as, at best, inconclusive and, at worst, incorrect (George 1988, 600). Instead, George focused on what he saw as a deeper issue: he claimed that Dummett's arguments in favor of intuitionism were ultimately circular, failing to establish a foundation independent of the very premises they sought to justify.

At the 1978 annual meeting of the International Institute of Philosophy, Dag Prawitz introduced his perspective on the philosophical dilemmas posed by intuitionistic logic (Prawitz 1980). Like George and Weir, Prawitz's work engaged with Dummett's intuitionist-based theory of meaning, but his philosophical interest in intuitionism derived from a broader context: the fundamental conflict between intuitionistic and classical logic and the need for a resolution. In Prawitz's view, our deductive practices implicitly take a stance on this conflict. Even attempts to reconcile the two logics often reveal a bias toward one over the other in actual reasoning. In the realm of philosophy, he argued, we should not only reflect on our deductive practices but also explicitly articulate what constitutes correct reasoning. The clash between classical and intuitionistic logic brings forth two intertwined questions: not only which logic is correct but also how one goes about justifying or challenging the validity of a logical law (Prawitz 1980, 1–2).

While Prawitz didn't advocate for a definitive solution to this philosophical clash, he argued against abandoning classical logic solely because it's challenging to construct a coherent theory of meaning for it. However, he also cautioned that clinging indefinitely to a practice that resists intelligible explanation reveals a level of irrationality. For Prawitz, this balance was essential: philosophy must strive for logical clarity without discarding established practices without substantial reason (Prawitz 1980, 10).

Beyond the criticism Dummett's work attracted, his interpretation of intuitionism sparked a wave of philosophical responses, prompting others to build alternative frameworks around Brouwer's ideas. Richard Tieszen, for one, saw Dummett's interpretation as limiting, arguing that Dummett had shifted focus away from key concepts in Brouwer's intuitionism, like intuition and consciousness, and instead emphasized meaning and language (Tieszen 1991). Motivated by what he viewed as critical gaps in Dummett's approach, Tieszen developed his own alternative theory of intuitionistic mental constructions, focusing on ideas in philosophical psychology and human intentionality (Tieszen 2000). Tieszen found inspiration in Arend Heyting's 1931 lecture, where Heyting described mathematical constructions as intentions that are either fulfilled or capable of being fulfilled (Heyting 1931). Following this line of thought, Tieszen explained how a theory of intentionality applied to mathematics can be used to deepen the philosophical interpretation of intuitionism while also advocating for it as a legitimate philosophy of mathematics (Tieszen 1994, 586).

Meanwhile, Neil Tennant notes how his interest in realism, the theory of meaning, and the concept of truth was provoked by Dummett's writings in the 1970s (Tennant 2002, vii), and yet he went on to claim that Dummett's argument, which incorporated the contemporary revival of anti-realism framed as an expanded form of intuitionism is "fallacious through and through" (Tennant 2002, 160). He developed his own argument for logical revision, in which he sought to redefine the principle of bivalence and Dummett's manifestation argument (Tennant 2002, 195–244). During the past four decades, Tennant devoted two book-length studies and several papers to his philosophical considerations on intuitionistic logic and intuitionistic theories (Tennant 1979, 1987a, b, 1994, 2002, 2005, 2009).

Carl Posy stands as another significant figure in philosophical discussions on intuitionism. Since 1974, when he published his first paper on Brouwer's constructivism (Posy 1974), up until 2020, when his textbook on mathematical intuitionism was released (Posy 2020), Posy's commitment to exploring intuitionism in both mathematical philosophy and logic has remained unwavering. Over the years, he has produced numerous works on the topic, from deep philosophical analyses of Brouwer's ideas to more accessible pieces aimed at introducing intuitionism to a wider audience (Posy 1976, 1992a, b, 1998, 2000, 2005, 2008). His 2020 monograph *Mathematical Intuitionism* was published in one of the Cambridge Elements series titled "Elements in the Philosophy of Mathematics".

Although the foundational debate within mathematics had abated and intuitionism no longer occupied a seat at the table, the philosophical engagement with intuitionism indicates otherwise; intuitionism, by all means, remains part and parcel of the philosophical discussion of mathematics. It was so in the past, as I have shown in the previous section, and it remains so today. As the publication of Posy's latest work clearly suggests, philosophers continue to consider intuitionism as an 'element' in the philosophy of mathematics. In the light of the many works surveyed throughout this chapter, one might even suggest that it was and remains a core element, in the sense that a comprehensive discussion of the philosophy of mathematics cannot afford not to include, or at least mention, Brouwer's intuitionism.

6.3 Concluding Remarks

Most scholarly works on intuitionism have typically focused either on how mathematicians responded to Brouwer's program or on how philosophers engaged with and expanded upon Brouwer's ideas. Yet, no one has systematically compared these two responses or investigated the reasons behind the stark contrast in their receptions. This chapter takes a first step toward filling that gap. It goes beyond simply proving the familiar adage that "one person's trash is another's treasure"; instead, it suggests that a theory's degree of acceptance can diverge dramatically between disciplines. The fact that Brouwer's intuitionism faded from mainstream mathematics nearly a century ago didn't diminish the keen interest it generated among philosophers of mathematics, logic, and language.

From a broader perspective, this chapter raises questions about the context in which Brouwer's intuitionism was discussed. A common claim is that Brouwer's program was too philosophical for practicing mathematicians, implying a distinction between foundational ideas that belong to the discipline of mathematics and those more properly situated within the philosophy of mathematics. Yet this distinction isn't always as clear-cut as it may seem. Some scholars, like Leon Horsten, have argued that the twentieth century has seen mathematicians engage deeply with theories that are, at their core, philosophical. According to this view, when mathematicians address foundational questions within their field, they are contributing to "foundational research," while philosophers who tackle the same questions are adding to the philosophy of mathematics (Horsten 2019). However, as Horsten points out and as was shown in this chapter, the boundary between the philosophy of mathematics and the foundations of mathematics remains vague.

Brouwer's intuitionism serves as an ideal example of this vagueness. Philosophers gravitated toward Brouwer's intuitionistic ideas while mathematicians were much less receptive, an aspect that highlights the core differences between how intuitionism was perceived by these two communities. By tracing the different responses to Brouwer's ideas, this chapter emphasizes philosophy's role as a discipline that not only reflects upon but also develops and complements scientific practices that were somewhat neglected. It also sheds light on the role scientific communities play in assessing and promoting new theories: each community adheres to its own standards, norms, and paradigms, which may account for why philosophers reacted to Brouwer's intuitionism so differently from mathematicians. Through this exploration, we see how the philosophical community's standards and openness to revisiting overlooked theories contributed to the preservation and evolution of intuitionism as a core element in the philosophy of mathematics, even as it receded from mathematical discussions.

References

Brown, James Robert. 2008. Philosophy of mathematics: A contemporary introduction to the world of proofs and pictures. In *Routledge contemporary introductions to philosophy*, 2nd ed. New York: Routledge.

Bueno, Otávio, and Øystein Linnebo, eds. 2009. *New waves in philosophy of mathematics*. Great Britain: Palgrave Macmillan.

Carnap, Rudolf. 1931. The logicist foundations of mathematics. In *Philosophy of mathematics: Selected readings*, ed. Hilary Putnam and Paul Benacerraf, 1st ed. Cambridge: Cambridge University Press.

Colyvan, Mark. 2012. *An introduction to the philosophy of mathematics*, Cambridge introductions to philosophy. Cambridge: Cambridge University Press.

Dawson, J. R., Jr. 1984. Discussion on the foundations of mathematics. *History and Philosophy of Logic* 5:111–129.

Dummett, Michael. 1975. The philosophical basis of intuitionistic logic. In *Truth and other enigmas*, 215–247. Cambridge: Harvard University Press.

Dummett, Michael. 1977. *Elements of intuitionism*. Oxford: Clarendon Press.

Dummett, Michael. 1978. *Truth and other enigmas*. London: Duckworth.

Dummett, Michael. 1991. *The logical basis of metaphysics*. Cambridge, MA: Harvard University Press.

George, Alexander. 1987. Reply to Weir on Dummett and intuitionism. *Mind* XCVI:404–406.

George, Alexander. 1988. Intuitionism, excluded middle and decidability: A response to Weir on Dummett. *Mind* 97:597–602.

George, Alexander. 1993. How not to refute realism. *The Journal of Philosophy* 90 (2):53–72.

George, Alexander, and Daniel Velleman. 2002. *Philosophies of mathematics*. Oxford: Blackwell.

Hesseling, Dennis E. 2003. *Gnomes in the fog: The reception of Brouwer's intuitionism in the 1920s*. Basel, Boston, Berlin: Birkhäuser Verlag.

Heyting, Arend. 1931. The intuitionist foundations of mathematics. In *Philosophy of mathematics: Selected readings*, ed. Hilary Putnam and Paul Benacerraf, 1st ed. Cambridge: Cambridge University Press.

Heyting, Arend. 1966. After thirty years. In *Studies in logic and the foundations of mathematics*, ed. Ernest Nagel, Patrick Suppes, and Alfred Tarski, vol. 44, 194–197. Elsevier. https://doi.org/10.1016/S0049-237X(09)70586-5.

Horsten, Leon. 2019. Philosophy of mathematics. In *The Stanford Encyclopedia of philosophy*, ed. Edward N. Zalta. Stanford University. https://plato.stanford.edu/archives/spr2019/entries/philosophy-mathematics.

Iemhoff, Rosalie. 2020. Intuitionism in the philosophy of mathematics. In *The Stanford Encyclopedia of philosophy (Winter 2019 Edition)*. Stanford University.

Israel, Giorgio, and Ana Millán Gasca. 2009. *The world as a mathematical game: John von Neumann and twentieth century science*, Science networks: Historical studies. Basel/Boston: Birkhäuser Verlag.

Körner, Stephan. 1962. *The philosophy of mathematics: An introductory essay*. New York: Harper Torchbook.

Leng, Mary. 2009. 'Algebraic' approaches to mathematics. In *New waves in philosophy of mathematics*, ed. Otávio Bueno and Øystein Linnebo, 117–134. Great Britain: Palgrave Macmillan.

Lindström, S., E. Palmgren, K. Segerberg, and V. Stoltenberg-Hansen, eds. 2009. *Logicism, intuitionism, and formalism: What has become of them?* Synthese library. Dordrecht, The Netherlands: Springer.

McCarty, D. C. 2005. Intuitionism in mathematics. In *The Oxford handbook of philosophy of mathematics and logic*, ed. Stewart Shapiro, 356–386. Oxford University Press.

Posy, Carl. 1974. Brouwer's constructivism. *Synthese* 27:125–159.

Posy, Carl. 1976. Varieties of indeterminacy in the theory of general choice sequences. *Journal of Philosophical Logic* 5:91–132.

Posy, Carl. 1992a. Kant's mathematical realism. In *Kant's philosophy of mathematics*, Studies in epistemology, logic, methodology, and philosophy of science, ed. Carl Posy, vol. 219. Dordrecht: Springer.

Posy, Carl. 1992b. Review: Dirk van Dalen, intuitionistic logic; Walter Felscher. Dialogues as a Foundation for Intuitionistic logic. *The Journal of Symbolic Logic* 57:754–756.

Posy, Carl. 1998. Brouwer versus Hilbert: 1907-1928. *Science in Context* 11 (2): 291–325.

Posy, Carl. 2000. Epistemology, ontology and the continuum. In *The growth of mathematical knowledge*, Studies in epistemology, logic, methodology, and philosophy of science, ed. E. Grosholz and H. Breger. Dordrecht: Springer.

Posy, Carl. 2005. Intuitionism and philosophy. In *The Oxford handbook of philosophy of mathematics and logic*, ed. Stewart Shapiro, 318–355. Oxford University Press.

Posy, Carl. 2008. Brouwerian infinity. In *One hundred years of intuitionism (1907–2007)*, ed. M. van Atten, P. Boldini, M. Bourdeau, and G. Heinzmann, 21–36. Birkhäuser Basel: Publications of the Henri Poincaré Archives.

Posy, Carl. 2020. *Mathematical intuitionism*, Elements in the philosophy of mathematics. Cambridge, MA: Cambridge University Press.

Prawitz, Dag. 1980. Intuitionistic logic: A philosophical challenge. In *Logic and philosophy*, ed. G. H. Von Wright and Dag Prawitz, 1–10. Dordrecht: Springer.

Putnam, Hilary, and Paul Benacerraf, eds. 1964. *Philosophy of mathematics: Selected readings*. 1st ed. Englewood Cliffs, NJ: Prentice-Hall.

Putnam, Hilary, and Paul Benacerraf, eds. 1984. *Philosophy of mathematics: Selected readings*. 2nd ed. Cambridge: Cambridge University Press.
Shapiro, Stewart. 2000. *Thinking about mathematics: The philosophy of mathematics*. Oxford: Oxford University Press.
Shapiro, Stewart, ed. 2005. *The Oxford handbook of philosophy of mathematics and logic*. New York: Oxford University Press.
Tennant, Neil. 1979. Language games and intuitionism. *Synthese* 42:297–314.
Tennant, Neil. 1987a. *Anti-realism and logic: Truth as eternal*, Clarendon library of logic and philosophy. Oxford University Press.
Tennant, Neil. 1987b. Natural deduction and sequent calculus for intuitionistic relevant logic. *Journal of Symbolic Logic* 52:665–690.
Tennant, Neil. 1994. Intuitionistic mathematics does not need ex Falso Quodlibet. *Topoi* 13:127–133.
Tennant, Neil. 2002. *The taming of the true*. Oxford: Oxford University Press.
Tennant, Neil. 2005. Contracting intuitionistic theories. *Studia Logica* 72:1–24.
Tennant, Neil. 2009. Natural logicism via the logic of orderly pairing. In *Logicism, intuitionism, and formalism: What has become of them?* ed. S. Lindström, E. Palmgren, K. Segerberg, and V. Stoltenberg-Hansen, 91–125. Netherlands: Springer.
Tieszen, Richard. 1991. What is a proof. In *Proof, logic and formalization*, ed. M. Detlefsen, 57–76. London: Routledge.
Tieszen, Richard. 1994. What is the philosophical basis of intuitionistic mathematics? In *Logic, methodology and philosophy of science IX*, ed. D. Prawitz, B. Skyrms, and D. Westersta, 579–594. Amsterdam: Elsevier.
Tieszen, Richard. 2000. Intuitionism, meaning theory and cognition. *History and Philosophy of Logic* 21 (3): 179–194. https://doi.org/10.1080/01445340051095793.
Troelstra, Anne Sjerp, and Dirk van Dalen. 1988. *Constructivism in mathematics*. Amsterdam: North-Holland.
von Neumann, John. 1931. The formalist foundations of mathematics. In *Philosophy of mathematics: Selected readings*, ed. Hilary Putnam and Paul Benacerraf, 1st ed. Cambridge: Cambridge University Press.
Weir, Alan. 1986. Dummett on meaning and classical logic. *Mind* 95:465–477.
Weir, Alan. 1988. Rejoinder to George. *Mind* XCVII:110–112.

Chapter 7
Interwoven Narratives: Unveiling the Social Fabric of Mathematics

Abstract This chapter examines the intricate ways in which human relationships, shared contexts, and philosophical tensions influence the evolution of mathematical thought. It focuses on the interplay of social identity, interpersonal conflicts, group contexts, and intellectual progress in mathematics through three key narratives. It first examines Hermann Weyl's transition from Göttingen to Princeton, highlighting how displacement disrupted his professional identity and creative focus. Next, it delves into Brouwer's intuitionism, emphasizing his efforts to balance solitary intellectual work with collective exchange, particularly through the Significs movement. Finally, the chapter analyzes the ideological rift and personal fallout between Brouwer and Hilbert, showing how their philosophical disagreements, shaped by interpersonal and historical tensions, influenced the direction of mathematics. Together, these stories reveal the critical role of social and psychological factors in shaping mathematical knowledge, offering a more nuanced picture of the evolution of intuitionism.

Keywords Social identity · Brouwer-Hilbert conflict · Significs movement · Interpersonal dynamics · Solipsism · Social history of mathematics · Mathematical agency · Philosophy of mathematical practice

In this chapter, I center on the social dimension of mathematics, exploring how identity, social belonging, and psychological challenges shape not just individual mathematicians but also the trajectory of the discipline itself. Throughout this book, figures like Weyl, Brouwer, Heyting, and Hilbert have emerged as intellectual giants, defined by their groundbreaking ideas and fierce debates. Here, however, I shift focus, examining the personal, social, and psychological forces that influenced their lives and work. At its core, this chapter seeks to uncover the intricate interplay between the individual and the collective in mathematics. While the discipline often celebrates solitary genius, the stories in this chapter remind us that mathematical innovation rarely happens in isolation. It is influenced by the communities' mathematicians are part of, the intellectual cultures they navigate, and the personal struggles they endure.

© The Author(s), under exclusive license to Springer Nature Switzerland AG 2025

K. Kish Bar-On, *Intuitionism Resocialized*, Studies in History and Philosophy of Science 66, https://doi.org/10.1007/978-3-032-05839-3_7

The story of Hermann Weyl, discussed in the first section, illustrates how professional identity and group belonging can shape a mathematician's trajectory. Weyl's move from Göttingen to Princeton in 1933, triggered by political upheaval and personal concerns, was more than a geographic shift. It disrupted the rich intellectual community that had nurtured his work, leaving him feeling disconnected and uncertain. This change not only affected his sense of self but also altered the focus of his mathematical contributions.

The second section explores the complex interplay between individual cognition and collective understanding in Brouwer's intuitionism. While Brouwer's philosophy, rooted in intuition and the self, is often seen as solipsistic, his engagement with the Significs movement reveals his acknowledgment of the communal aspects of knowledge creation. Through this engagement, Brouwer emphasized how individual ideas gain vitality and broader influence when shared and shaped within a collective. The section sets the stage for exploring how these dynamics evolve into broader interpersonal and ideological conflicts, as seen in Brouwer's later professional life.

In the third section I home in on the dramatic fallout between Hilbert and Brouwer, a conflict that intertwined personal animosities with profound philosophical disagreements. The dispute culminated in Brouwer's removal from the editorial board of the prestigious *Mathematische Annalen*, marking a turning point in his career and in the trajectory of intuitionistic mathematics. Compounded by the broader political upheaval of the time, including rising nationalism and tensions within academia, the conflict highlighted how personal rivalries and external pressures could profoundly impact intellectual movements.

These three cases, taken together, illustrate how mathematics is not only shaped by ideas and proofs but also by personal convictions, institutional dynamics, and broader cultural tensions. Each section, in its own way, reveals how the social world enters into mathematical reasoning and practice. Building on these case studies, the last section turns to a more explicit engagement with the philosophy of mathematical practice (PMP). I survey some of the central arguments within this tradition, showing how they intersect with the themes explored throughout the book: personal relationships, social structures, and psychological dynamics. I also highlight where these accounts remain incomplete when considered in light of the preceding case studies, particularly in their treatment of related psychological theories and the integration of complex social interactions into mathematical practice.

7.1 Identity, Belonging, and the Challenges of Change

7.1.1 Hermann Weyl: From Philosophical Physics to His Move to Princeton

Up until now we have carefully examined the different philosophical influences on Weyl's view regarding the foundations of mathematics, but his work in physics was also influenced by different philosophical approaches. Several scholars have discussed the relationship between Weyl's works in physics and Husserlian phenomenology. Thomas Ryckman (2003) claims that the transcendental phenomenological idealism of Husserl is the philosophical motivation underlying Weyl's a priori posit of the concept of local gauge symmetry and his subsequent prolonged defense of it in *Space-Time-Matter* (Weyl 1918). While Ryckman maintained that Weyl's intellectual exchange with Husserl's philosophy never ended (Ryckman 2005, 115), Richard Feist argues that it is not entirely clear to which aspects of Husserl thought Weyl committed himself (Feist 2004, 156–58).

Whereas some scholars suggested that Weyl was inspired by Kant's transcendental idealism and by the writings of Leibniz (Friedman 1999; Newman 1957), others show that Leibniz, and specifically Kant, were marginal sources of inspiration in Weyl's philosophy of physics and on his work on unified field theory as well as on his problem of space (Eckes 2018). Christophe Eckes describes the evolution of Weyl's philosophy of physics and distinguishes between Weyl's early "apriorism" (1918–1921) and his later refinement of it (1921–1923). According to Eckes, while in 1918 Weyl held a mathematical definition of space as a "form of phenomena" that is mostly Kantian, in 1923 Weyl redefined the expression "form of phenomena" to mean "the set of all properties characterizing the a priori essence of space" (Weyl 1922, 265)—a definition that is not Kantian at all (Eckes 2018, 172).

Philosopher Norman Sieroka points to another influential figure in Weyl's thought, the German Idealist Johann Gottlieb Fichte, and highlights several Fichtean themes in Weyl's work (Sieroka 2007). Besides focusing on the connection between Weyl and Fichte, Sieroka draws our attention to another question: why would Weyl, who was a mathematician and a political liberal, would ever read Fichte, who was considered "mathematically ignorant" and his work was prominent in a national conservative, practical philosophy? The answer lies in the interaction between Weyl and his friend and colleague Fritz Medicus, a philosopher and Fichte scholar, who read him in a particular and liberal way.

Sieroka's paper points to an important epistemological and social aspect that must be taken into account when analyzing the impacts and interactions between scientists and ideas:

> [...] certain texts and documents cannot be fully understood on their own (as is here the case with the Fichtean themes in Weyl's writings), these texts or documents are investigated as part of a constellation consisting of additional persons, theories, and documents. In the present case, attention to a wider constellation also including political considerations will obviously be needed to enable a distinction between the reception of Fichte's work by Weyl

> and Medicus in Switzerland (the 'Zurich Fichte') and the national conservative reception of Fichte's work in most parts of Germany. Here it is clear that a politically indifferent methodological approach like that common in History of Ideas will not be helpful. Konstellationsforschung also differs from other methodological approaches such as, for example, Discourse Analysis, by placing special weight on the creativity of the actors in a constellation, a creativity which often reveals itself in intense intellectual exchanges over rather short time-spans (as was the case with Medicus and Weyl reading Fichte together). Those actors are assumed to have a sphere of possibilities available to them within a specific intellectual and conceptual space (Denkraum). (Sieroka 2007, 85–86)

As I have shown in Chap. 5, the broader constellation and the political considerations have played a significant role in the evolution of Weyl's mathematical views. But Sieroka points here to a specific social-epistemic approach; by describing the practitioners' "sphere of possibilities" as dependent on their "intellectual and conceptual space (Denkraum)," Sieroka refers here, in a footnote, to Ludwik Fleck's notions of "thought style" and "thought collective" (Sieroka 2007, 86, ft. 6). While Sieroka does not develop the idea beyond this footnote, I would like to take a closer look at the connection between Weyl's philosophical influencers and Fleck's theory of thought styles.

Ludwik Fleck was a Polish-Jewish microbiologist, who developed in the 1930s the first system of historical philosophy and sociology of science (Sady 2021). According to Fleck, cognition is not an individual process of a particular consciousness but the result of social activity since "the existing stock of knowledge exceeds the range available to any one individual" (Fleck 1981, 38). As such, Fleck highlights that science is not made by individuals working alone in a vacuum. Rather, it is a social, collective activity:

> A truly isolated investigator is impossible, and so also is an a-historic discovery, or a styleless observation. An isolated investigator without bias and tradition, without forces of mental society acting upon him, and without the effect of the evolution of that society, would be blind and senseless. Thinking is a collective activity, just as choral singing or conversation. It is subjected to specific changes in time, and displays a historic continuity of these changes. Its product is a certain picture, which is visible only to anybody who takes part in this social activity, or a thought which is also clear to the members of the collective only. What we do think and how we do see depends on the thought-collective to which we belong. (Fleck 1986, 77)

Fleck defines a thought collective as a community of people mutually exchanging ideas or maintaining intellectual interaction (Fleck 1981, 39), in the sense that members of that collective not only adopt specific ways of perceiving and thinking but also continually transform them. This transformation does not occur only in their minds, but it is part and parcel of their interpersonal space, social environment, and collective culture. We can easily spot such interactions in everyday life when individuals in a group context start to discuss ideas that give rise to new ways of thinking that would not have evolved without the group context. That is how thought styles arise, and with them also arises a collective mood, which is a form of thought structure that does not belong to any particular individuals but is a characteristic of the group and inclines group members to act in a certain way (Fleck 1981, 44).

A thought collective exists wherever two or more people are exchanging thoughts. If a new person joins in, a new collective arises. The previous mood will dissolve and, with it, the special creative force of the former small collective. Some collectives last shortly, even only as long as an individual conversation or a meeting last, while others can last years. We can think of at least two events in Weyl's life that can be examined through the collectively oriented perspective Fleck provides. The first is Weyl's meeting with Brouwer in Engadin in 1919, and the second is Weyl's move from Germany to Princeton in 1933. The meeting in Engadin is considered a short-term thought collective, but it had apparent implications for the scientific development of mathematical intuitionism, exposing it to wider audience while creating two versions of intuition, Brouwer's intuitionism and Weyl's intuitionism. The move to Princeton, however, reveals a different aspect of Weyl's professional trajectory, intersecting collectivity with individuality.

In 1933 Weyl accepted an offer to move from his position as Hilbert's successor in Göttingen to the Institute for Advanced Study, which was then newly founded in Princeton. Most historians agree that the reasons behind Weyl's decision to leave Germany were a combination of the threat of coming political events, the dismissal of influential mathematicians from the mathematical centers in Berlin and Göttingen, and his wife being Jewish (Sigurdsson 1991; Newman 1957; Atiyah 2003). In 1930, Weyl returned to Göttingen after a peaceful and prolific seventeen years at the Swiss Federal Institute of Technology (ETH) in Zurich, and he explained his return with the following words:

> I have returned because I did not want to miss the contact with youth without which old age fades away into rigid loneliness, because I recognize that intellectual community is the proper site of scientific growth, and hence I acknowledge it as a serious duty to cultivate the tradition of science, and finally to embed myself again in the national community. (Weyl 1930, 653–54)

Weyl's words echo Fleck's principles for making scientific progress as part of a thought collective. He was keen on continuing his work in Göttingen and fostering an intellectual community of students and scholars. However, the political and economic situation in Germany turned things around, and in October 1933, Weyl immigrated to the United States with his family. Weyl's move to the rising center of mathematical research consolidated his research goals, as he concentrated on mathematics and turned away from physics and philosophy (Sigurdsson 1996).

Yet this move took a toll from Weyl, personally and professionally. Historian Skúli Sigurdsson compares the intellectual migrations of Weyl, Max Born, and Erwin Schrodinger from Europe to the United States and shows that changes in the cultural, social, and intellectual environments had a noticeable impact on the direction their research trajectories took. Born and Schrodinger continued their old projects, started new ones, and "cast their philosophical nets ever wider" (Sigurdsson 1996, 49). On the other hand, Weyl arrived in the states spiritually and physically exhausted and felt he was in "danger of losing myself".[1] He concentrated his work

[1] Hermann Weyl to Max Born, Princeton, N.J., Dec. 9, 1942, Max Born papers 837, SPKB.

efforts on relatively pure mathematics, and most of his publications during his time at Princeton revolved almost exclusively around mathematical topics (for example, *Classical Groups* (Weyl 1939), *Meromorphic Functions and Analytic Curves* (Weyl 1944), *Symmetry* (Weyl 1952), and *Algebraic Theory of Numbers* (Weyl 1954)). Weyl emphasized the break with the past and his turn away from physics when he told a biographer of Eddington's in 1953 that "it is now more than 20 years that I ceased to follow the development of theoretical physics and the various speculative attempts at establishing a unified theory".[2] While this statement was not entirely true, as Weyl did engage with physics to some extent throughout these 20 years, he did not pursue any new research directions in theoretical physics after moving to Princeton.

By joining the Institute for Advanced Study, Weyl became part of a thought collective that was different from his own professional group in Göttingen or Zurich; the institute in Princeton was much more mathematically oriented, and it came across in its members' research interests, their working frameworks, and norms of inquiry. Hence it is quite plausible that Weyl had shifted his research direction to align with his new colleagues. What makes it even more reasonable is that among his new colleagues at Princeton was a familiar face from Germany—the mathematician Emmy Noether. Noether and Weyl were colleagues in Göttingen between 1931–32, yet they first met much earlier, in 1915, and have been members of the same mathematical circles (specifically those of Hilbert and Helmut Hasse) ever since (Roquette 2008). Weyl was impressed by Noether's originality and creativity in mathematics and physics, and it has been argued that Weyl's work on gauge symmetry was an application of Noether's theorem of 'local' symmetries (Brading 2002). While Noether and Weyl had a fair share of disagreements about axiomatic methods and the proper way of securing the logical truth of mathematical results, by 1935, in his obituary of her, Weyl acknowledged that axiomatization had become a powerful tool of concrete mathematical research itself, in particular, due to Noether's contributions (Weyl 1935, 209).

Their close friendship and Weyl's high esteem of Noether may have also contributed to the redirection of Weyl's research interests during his time at Princeton. Here, we can also think of Noether as part of Weyl's thought collective, as Weyl's personal and academic interactions with Noether and her work shaped his research trajectories, which shifted specifically towards algebraic theories, groups and symmetry from 1935 onwards.

The idea that collective norms, social bonds, and personal interactions affect mathematicians' work trajectory is not surprising or new; it has been extensively discussed since the 1970s in philosophy of mathematical practice and sociology of science, and I will come back to these discussions in Sect. 7.4. What I wish to highlight in Weyl's story is the importance of exploring the *social psychology* of

[2] Hermann Weyl to A. Vibert Douglas, Zurich, Oct. 31, 1953, Hermann Weyl papers HS 91:173, ELAZ.

mathematicians: how social identification and group belonging can shape not only a mathematician's sense of self but also their scientific output and career direction.

7.1.2 *The Intersection of Social Identity and Mathematics*

The history of mathematics is a story about the evolution of knowledge as well as the evolution of mathematicians as social creatures: their identities, their interactions with others, and the ways in which their social and cultural contexts shape their professional trajectories and the creation of new theories. Claims about the connection between mathematical knowledge and social factors have been around since the mid-twentieth century. Social constructionist approaches to mathematics hold that mathematical objects and structures are not discovered but are instead brought into being through the collective practices of mathematicians. Some scholars investigate the ways in which mathematical progress is shaped by broader political and social conditions, as well as by the roles of specific individuals within these contexts (Bloor 1991; MacKenzie 1981; Mehrtens 1987, 1990). Others go further, arguing that the very content of mathematics is inherently social in nature (Ernest 1998; Hersh 1997; Ernest 1992). More recent contributions to this discourse compare socially grounded accounts of mathematics with traditional ontological positions, including platonism, fictionalism, and modal nominalism.[3] Historical studies have further contributed to this discussion by illustrating how mathematical practices are shaped by the social organization of research communities, emphasizing that certain proofs were not the product of isolated reasoning but of sustained collective effort, shaped by face-to-face communication, informal pedagogy, institutional support, and shared norms (Steingart 2012).

Mathematicians are therefore part of collaborative networks, professional communities, and institutional cultures that shape their work and, in turn, their social identity. The concept of social identity refers to an individual's sense of belonging to particular groups, shaped by their experiences, values, and relationships (Turner and Reynolds 2010). Social identity theory—one of the leading behavioral theories in social psychology—was developed by social psychologists Henri Tajfel and John Turner, and it emphasizes how individuals derive a sense of who they are based on the groups to which they belong (Tajfel et al. 1979; Tajfel 1981; Turner 1975; Turner and Onorato 1999). These groups, whether cultural, professional, or institutional, provide frameworks for understanding one's role in the world, fostering a sense of belonging, and shaping one's values and goals. For professionals, especially those in highly specialized fields like mathematics and science, this intertwines with

[3] For example, Julian Cole's theory of "practice-dependent realism" proposes that mathematical domains emerge from socially structured practices (Cole 2009, 2013, 2015). Jill Dieterle has challenged this view, contending that it fails to account for the timelessness of mathematical entities (Dieterle 2010), while Shay Allen Logan has developed a hybrid model that combines elements of fictionalism with social constructivism (Logan 2015).

professional identity, which encompasses how they see themselves in relation to their discipline, their peers, and their intellectual endeavors. Together, social identity and professional identity form the bedrock of how individuals navigate their lives, relationships, and careers, and how they respond to challenges, changes, and opportunities.

For Weyl, these identities were inextricably linked. From a psychological perspective, social identity influences how a mathematician views his place within a group: whether as a pioneer, collaborator, or a student of tradition. Professional identity, on the other hand, governs his self-perception as a contributor to mathematical knowledge, his goals within the discipline, and his adherence to the norms and standards of the field. Weyl's early career in Göttingen, where he succeeded David Hilbert, placed him at the epicenter of mathematical innovation. This position was not merely a professional achievement but also a social one—it solidified his status as part of an intellectual community that was redefining mathematics and physics in the early twentieth century. Weyl's identity was rooted in this environment, which provided intellectual stimulation, a sense of belonging, and an affirmation of his contributions. It was this strong sense of group belonging that made Weyl's endorsement of Brouwer's ideas not only surprising but also felt by some as a betrayal of Hilbert's legacy. To this extent, Weyl was a student of tradition, until he wasn't.

Social identity within academic communities is not only about professional status but also about the interpersonal relationships and shared values that define those communities. Researchers in social psychology have shown that individuals derive a significant portion of their self-esteem and sense of purpose from their inclusion in valued groups (Lambert et al. 2013; Hogg et al. 2017; Abrams et al. 1990; Bergami and Bagozzi 2000). In academic settings, this often means being part of a network of mentors, colleagues, and students who share common intellectual goals. These communities foster creativity, collaboration, and the exchange of ideas, all of which are essential for scientific and mathematical progress.

Weyl's move from Göttingen to Princeton disrupted this dynamic. In Göttingen, he was part of a tightly knit group of mathematicians and physicists, including figures like Emmy Noether and David Hilbert, who shared a common language and intellectual tradition. His professional identity was reinforced by his role within this community, where he was both a student and a leader. At Princeton, however, Weyl found himself in a different cultural and institutional setting, where he struggled to establish the same sense of belonging. This lack of connection had profound implications for his professional identity and, consequently, his scientific focus.

Professional identity is often seen as stable, but it is, in fact, highly susceptible to changes in one's social and institutional environment. Moving to a new institution, adopting a new cultural context, or navigating political upheaval can challenge an individual's sense of self within their profession. For mathematicians, whose work often relies on deep, sustained intellectual engagement, these disruptions can be particularly destabilizing.

In Weyl's case, his professional identity was not only tied to his intellectual achievements but also to his role within specific academic communities. At

Göttingen, his identity was shaped by his status as Hilbert's successor and his contributions to foundational debates in mathematics and physics. The political turmoil of the 1930s and his subsequent move to Princeton disrupted this identity. Although Princeton was a prestigious institution, it lacked the established intellectual traditions of Göttingen, where Weyl was familiar with the norms and social hierarchy. Weyl's sense of isolation in this new environment led him to shift his focus from the cutting-edge intersections of mathematics and physics, which were his main focus at Göttingen, to more traditional areas of pure mathematics, following his colleagues at Princeton.

Relocating to a new cultural or institutional setting often involves a renegotiation of identity. Social psychologists have found that moving to a new environment can lead to identity conflict, where individuals must reconcile their past affiliations with the demands and expectations of their new context (Gibson 2001; Berry 1980; Tadmor et al. 2009). This process, known as *acculturation*, can be particularly challenging for professionals whose identities are deeply tied to specific cultural and institutional traditions.

Weyl's move to Princeton was not just a geographic relocation but also a cultural and professional dislocation. At Göttingen, he was immersed in a European intellectual tradition that valued interdisciplinary collaboration and rigorous debate. In Princeton, he encountered a nascent academic culture that lacked the same depth of tradition. Moreover, his sense of disconnection from the physicists' community at Princeton—a field where he had been a central figure—exacerbated his feelings of isolation. His poignant admission that he does not belong with the real physicists highlights the psychological toll of displacement and its impact on his professional identity.

Weyl's story underscores the importance of social and professional identity in shaping scientific trajectories. His experience illustrates how cultural and social settings influence not only individual careers but also the trajectory of a discipline. Weyl's sense of isolation in Princeton, juxtaposed with his earlier belonging to tightly knit intellectual circles, highlights the delicate balance between individuals and the collectives they are part of. This theme finds a fascinating counterpart in Brouwer's views on the role of individuals and groups in mathematical thinking, that will be explored in the next section.

7.2 Brouwer and the Significs: Between Individual Cognition and Collective Understanding

Brouwer's intuitionism, with its central premise that mathematics is a creation of the mind, is deeply rooted in an individualistic perspective. For Brouwer, intuition serves as the sole foundation for certain knowledge, with the Self acting as the wellspring of all understanding. In this view, any epistemic claim must derive its existence through the Self, leading to the frequent association of intuitionism with solipsism—the philosophical stance that knowledge beyond one's own mind is inherently uncertain (Blum 2005; van Stigt 1982).

Yet, Brouwer's active engagement with the Significs movement, a sociolinguistic and philosophical circle that thrived in the early twentieth century, presents a more nuanced picture of his philosophy. His participation in this movement underscores his acknowledgment of the communal aspects of knowledge and the significance of group thinking, and casts doubt on the simplistic association between intuitionism and solipsism (Franchella 1995; Dummett 1975; Hesseling 2003). Moreover, it hints at Brouwer's broader vision: a belief in grounding mathematics within the tangible, social realities of human life. Brouwer describes the Significs movement as "intuitive significs," explaining that it "concerns itself with the creation of new words forming a new code of elementary means of communication" (Brouwer 1918). His wider vision for the reform of significs extends into the political sphere and emphasizes collective modes of communication, grounded in scholars' capacity to collaborate and engage in shared intellectual activity.

The concept of "Significs" originated with Victoria Lady Welby, a self-taught British philosopher who introduced the term in her 1896 paper, "Sense, Meaning and Interpretation" (Welby 1896). She developed an approach that treated the act of communication as a reciprocal process, giving equal weight to the speaker or writer's use of signs and the listener or reader's interpretation of them. In doing so, she moved away from the dominant philological-historical traditions of her time, which tended to privilege textual origin over reception (Petrilli 2009). Central to her thinking was the belief that a refined theory of signs could significantly enhance human communication. She was deeply convinced that a fuller understanding of how people exchange meaning could not only improve dialogue but also serve as a foundation for addressing (and potentially resolving) many of the social problems facing modern society (Schmitz 1990).

The idea of Significs was brought to the Netherlands in 1908 by Frederik Willem van Eeden, though he did not explicitly credit Lady Welby (van Eeden 1908). A few years later, in 1912, Jacob Israel de Haan further introduced Welby's theory in his paper "New Philosophy of Legal Language" (de Haan 1912). This laid the groundwork for the formation of a dedicated group. In 1917, van Eeden, de Haan, Gerrit Mannoury, and Brouwer came together to establish the "International Academy for Practical Philosophy and Sociology," which included prominent figures such as Henri Borel, physicist Leonard Ornstein, and H. P. J. Bloemers (Brouwer 1946, 465).

By 1922, the group underwent significant changes. After de Haan departed and linguist-theologian Jacques van Ginneken joined, the organization rebranded itself as the "Signific Circle" ("Significsche Kring"). With the formation of the Significs Circle, the group expanded its focus beyond linguistic critique and synthesis. Its revised aim was to investigate the relationship between language and the inner needs and inclinations of the human soul, with the broader intention of shaping future social and psychological conditions (Brouwer 1946, 468). The circle members met regularly for four years, but was disbanded in 1926, driven by van Eeden's declining health and van Ginneken's retirement, marking the end of an era for this interdisciplinary and philosophical initiative.

The ultimate aim of the Academy and the Significs movement, as envisioned by Brouwer, was to explore how individual ideas could transcend their solitary origins to become part of the collective consciousness, thereby amplifying their societal influence. For Brouwer, this process hinged on the power of shared human understanding, as he believed that when a group of independent thinkers, guided by refined and genuine human sensibilities, worked together on a common task, their mutual understanding would naturally generate a shared language (Brouwer 1946, 466). This language would serve as a medium for their ideas to resonate not just within their intimate circle but across wider social contexts, facilitating broader societal understanding and impact.

Brouwer articulated this principle by emphasizing the transformative potential of collective thought. He described a thought as an "embryonic deed," suggesting its inherent potential to evolve and influence (Brouwer 1946, 466). However, he argued that this potential was significantly amplified when the thought became a shared, intimate conviction within a group of people rather than remaining confined to the mind of a solitary individual. In Brouwer's view, the collective conviction of a group, united by mutual understanding and a shared sense of purpose, imbued an idea with greater vitality and the capacity to enact meaningful change in the broader world.

Brouwer's deliberate phrasing, particularly his emphasis on scientists' thoughts being "formed in the mutual understanding" of their groups, reveals his acknowledgment of the distinction between individual and collective work, as well as the advantages of the latter. He implies that when ideas emerge from a group of individuals, they gain access to the "mutual understanding" of the broader community more effectively than isolated, individual efforts can achieve. In essence, the act of collective thinking shapes ideas in a way that makes them inherently sharable, transforming them into products of a collective mind.

This collaborative genesis allows ideas to evolve, flourish, and find resonance far more effectively than when confined to the thoughts of a single person. By highlighting this dynamic, Brouwer underscores the role of the group, the community, or the collective as the critical factor in the creation and dissemination of knowledge. Remarkably, this perspective aligns with the idea that shared intellectual labor is central to progress—a notion that contrasts sharply with the solipsistic tendencies often associated with Brouwer's philosophical positions. His insights here suggest a nuanced appreciation for the social dimensions of intellectual activity, reinforcing the importance of collaboration in advancing understanding and innovation.

But Brouwer did not arrive to these insights merely by himself; he was deeply influenced by his teacher and mentor, Gerrit Mannoury. Mannoury is regarded as one of the key figures who shaped the development of Brouwer's philosophical and mathematical thought (van Atten 2020; van Dalen 2013; Hesseling 2003; van Stigt 1990). Brouwer himself credited Mannoury with fundamentally transforming his view of mathematics, describing him as a uniquely insightful thinker whose writings and conversations revealed to Brouwer a deeper, more dynamic understanding of mathematical thought (Brouwer 1946, 192–93). The teacher-student relationship quickly evolved into a "dialectic partnership," and Mannoury became one of

Brouwer's few lifelong friends. Their close bond and ongoing intellectual exchange gave rise to rich discussions on topics they both deeply cared about, including language, philosophy, symbolic logic, conceptual analysis, and, most notably, significs (Kirkels 2013, 2019).

Mannoury was a remarkable figure—a mathematician without a formal degree, a committed Marxist, and an outspoken political activist. His rebellious spirit led him to challenge established paradigms and question authority, believing that language held the key to understanding human behavior and mathematics (Mannoury 1917; Meertens 1956). In particular, he viewed mathematics as a practical and socially significant science, whose role is to liberate thought from entrenched dogmas like determinism and rigid rationalism, and argued that true scientific progress requires recognizing that even the most firmly held mathematical beliefs are open to revision (Mannoury 1917, 18–19).

In one of their documented dialogues during a meeting of the Significs circle, Brouwer and Mannoury shared how each of them views mathematical practices (Brouwer 1937). Brouwer argued that mathematics inherently possesses significance, as it emerged from humanity's practical need to navigate life's complexities. He saw mathematical evolution as intertwined with the development of a corresponding language, serving both human necessity and societal function. Mannoury, in contrast, suggested that much of mathematics had become detached from these roots, cultivated instead as "art for art's sake," with no clear practical aims. Brouwer firmly disagreed, maintaining that even the most abstract or recreational branches of mathematics retain indirect connections to societal needs. He described mathematical thought as a network that links valuable aspects of life, with its importance rooted in practical and social relevance even when these connections are not immediately obvious (Brouwer 1975, 450–51).

The dialogue between Brouwer and Mannoury highlights Brouwer's conviction that mathematics is deeply tied to social and practical contexts, even as Mannoury questions its detachment and aesthetic independence. Brouwer firmly believed that mathematics does not exist or evolve in isolation from societal circumstances; he believed that the true significance of a mathematical formula lies in its "asserting character," which is realized through its active manipulation or operation within a community (Brouwer 1975, 451; Placek 1999). This emphasis underscores the importance of human interaction in giving mathematical constructs their meaning and value. Echoing his broader ideas on the role of group thinking, Brouwer's exchange with Mannoury reinforces the idea that the community plays a crucial role in shaping and legitimizing mathematical knowledge.

To illustrate the communal and manipulative nature of mathematical formulas, Brouwer draws an analogy to a bank telegram and bookkeeping, connecting practical communication with the concept of mathematical truth (Brouwer 1975, 451). The act of sending a telegram represents an intentional, goal-driven use of language—what Brouwer terms "living, asserting language." The codewords within the telegram act as links, bridging immediate actions with broader, ultimate goals. This highlights how communication is inherently tied to purpose and how language serves as a medium to navigate between means and ends. When the content of the telegram transitions into bookkeeping, it undergoes a transformation, and the explicit connection between the immediate actions and their larger goals becomes obscured. For the bookkeeper, this transition dissolves into the general feeling of duty fulfilled, paralleling the mathematician's experience of truth as a deeply emotional phenomenon tied to abstraction and process.

This analogy underscores Brouwer's belief that mathematical truth is not solely an abstract, isolated construct but is instead deeply tied to its use and manipulation within a community. Much like the banker's telegram, mathematical expressions derive their significance from their operational context and their function as connectors between intentions and outcomes. This perspective reinforces the communal aspect of mathematics: just as the bookkeeper participates in the bank's broader communication network, mathematicians engage with formulas and expressions within a shared intellectual and social framework.

Brouwer's emphasis on the emotional dimension (i.e., how the bookkeeper feels fulfillment akin to the mathematician's perception of truth) connects to his broader argument that mathematical thought is a human activity deeply embedded in societal needs and contexts. This stands in contrast to his so-called solipsistic views, or any views of mathematics as purely individualistic or disconnected from practical realities. For Brouwer, the shared processes of using, manipulating, and interpreting mathematical formulas anchor their truth, making the communal and societal context of mathematics indispensable to its understanding and development.

In this light, Brouwer's notion of truth transcends formal systems, challenging the idea that mathematical rigor resides solely in symbolic representation. Truth, for Brouwer, cannot be fully reduced to formulas or strict rules without losing its deeper connection to human experience and societal relevance (Brouwer 1975, 451–52). Just as the bookkeeper's entries fail to capture the full scope of a banker's intentions and strategies, mathematical systems cannot fully encapsulate the emotional and practical dimensions of truth. This analogy reinforces Brouwer's belief that both mathematical and practical truths are deeply rooted in their social and communal origins, further emphasizing the interconnectedness between individual cognition and collective understanding.

The communal threads woven into Brouwer's thoughts on the social context of mathematics stand in striking contrast to his solipsistic views. While his early claims often leaned toward the isolation of the self, asserting that the existence of other minds could not be proven, his later reflections revealed a more nuanced perspective. In some of these later works, Brouwer returned to solipsistic statements, reiterating doubts about the existence of other minds (Brouwer 1948a). Yet, simultaneously, his engagement with the broader mathematical community seemed to suggest otherwise. As Heyting observed, Brouwer at times described mathematics as a collective activity: a pursuit of the mathematical community as a whole (Heyting 1978).

This description is encapsulated in Brouwer's theory of the creating subject (Brouwer 1948a; van Atten 2017; Troelstra and van Dalen 1988; Kreisel 1967; Troelstra 1969; van Atten 2004). The theory proposes that the knowledge held by the creating subject is not isolated to the individual but reflects the cumulative knowledge of the entire community. According to this view, the community is not merely a passive repository of mathematical truths but an active force that uses, shapes, and creates knowledge (Posy 2020; Niekus 2010). In essence, the community becomes an engine of mathematical evolution, embodying a collective consciousness that transcends individual cognition.

Heyting, too, offered his own insights into the dynamic interplay between individuals and their communities. He emphasized that the individual is inseparable from the culture and society in which they reside, and maintained that the similarity and social connection between individuals lead to mutual perception and to the growth of knowledge. His own conviction in Japan's existence, for example, was not derived from personal experience but from education and a shared belief in the intersubjective nature of perception (Heyting 1978). Such convictions, so he claimed, are formed through intersubjectivity—a shared understanding that binds individuals together and fosters collective knowledge.

Through these perspectives, both Brouwer and Heyting highlighted the tension between the solitary and the communal in the world of mathematical thought. While Brouwer's solipsistic tendencies suggested an inward focus, his acknowledgment of the communal aspects of mathematical activity revealed an undeniable appreciation for the social dimensions of knowledge creation, bridging individual cognition with the shared understanding of a broader collective.

The interplay between solitary intuition and group dynamics might have shaped Brouwer's intellectual contributions, but it also foreshadowed the fractious debates he would face on a much larger stage. As Brouwer's personal and philosophical convictions collided with Hilbert's views, these tensions spilled over from the realm of ideas into the deeply personal and politically charged disputes that would reshape the landscape of mathematics. Let us now turn to explore the fallout from the Brouwer-Hilbert battle, examining how philosophical disagreements can evolve into personal rifts with lasting consequences for individuals and their communities.

7.3 Hilbert vs. Brouwer: Ideological Battles and Personal Fallouts

The fallout between Brouwer and Hilbert, briefly touched upon in this book's introduction, remains one of the most fascinating and pivotal episodes in the history of mathematics—a clash where personal and political tensions intertwined with intellectual rivalries. This dramatic turn of events not only shaped Brouwer's career but also left a profound mark on the evolution of intuitionistic mathematics. To start at the conclusion: by 1928, Brouwer had served as an editorial board member of the *Mathematische Annalen* for fourteen years. At the time, the *Annalen* was the most prestigious mathematical journal in the world, and securing a place on its editorial team was a significant professional honor. Yet, by 1929, Brouwer was no longer part of this elite group. The catalyst for his removal was no other than David Hilbert, another towering figure in mathematics and a fellow editor of the journal.

In October 1928, Hilbert sent Brouwer a short note:

> Dear Colleague,
>
> Because it is not possible for me to cooperate with you, given the incompatibility of our views on fundamental matters, I have asked the members of the board of managing editors of the *Mathematische Annalen* for the authorization, which was given to me by Blumenthal and Carathéodory, to inform you that henceforth we will forgo your cooperation in the editing of the Annalen and thus delete your name from the title page. And at the same time I thank you in the name of the editors of the Annalen for your past activities in the interest of our journal.
>
> Respectfully yours,
>
> D. Hilbert. (van Dalen 1990, 18)

At the time Hilbert sent his letter to Brouwer, *Mathematische Annalen* boasted an editorial team composed of some of the most illustrious figures in mathematics and physics. Names like Albert Einstein, Otto Blumenthal, Constantin Carathéodory, and David Hilbert himself graced the journal's masthead, alongside a roster of distinguished collaborators that included Ludwig Bieberbach, Harald Bohr, Richard Courant, Walther von Dyck, Arnold Sommerfeld, and Otto Hölder. Each of these individuals carried significant weight in their respective fields, contributing to the journal's reputation as the pinnacle of mathematical scholarship. While the daily operations of the *Annalen* were ostensibly overseen by Otto Blumenthal, it was Hilbert who held the ultimate authority.

The story of Brouwer and Hilbert's relationship is one of gradual transformation, from mutual admiration to the point of conflict. It began with a foundation of respect, especially from the young Brouwer, who in 1909 referred to Hilbert as "the first mathematician of the world." Their mutual regard seemed genuine in the early years, as evidenced by Hilbert's actions: not only did he write Brouwer a letter of recommendation for a position at the University of Amsterdam, but in 1919, he even went so far as to offer Brouwer a prestigious chair in Göttingen—a remarkable gesture that Brouwer ultimately declined.

However, the amicable dynamic between the two began to shift during the 1920s as the foundational debates in mathematics gained momentum. These debates highlighted philosophical divides, particularly between Brouwer's intuitionism and Hilbert's formalism. Brouwer's lectures in Berlin and Vienna in 1927 and 1928 served as a turning point. In these talks, Brouwer openly challenged Hilbert's formalist approach, signaling a departure from respectful disagreement to outright opposition. What began as an intellectual rift now took on a personal dimension, setting the stage for the escalating tensions that would culminate in Brouwer's removal from the *Mathematische Annalen* editorial board.

To remove Brouwer from his position as an associate editor of *Mathematische Annalen*, Hilbert needed the backing of the other chief editors: Felix Klein, Albert Einstein, and Otto Blumenthal. In a decisive letter to Einstein, Hilbert articulated his unyielding refusal to remain on the editorial board alongside Brouwer:

> Just to forestall misunderstandings and further ado, which are totally superfluous under the present circumstances, I would like to point out that my decision—to belong under no circumstances to the same board of editors as Brouwer—is firm and unalterable. To explain my request I would like to put forward, briefly, the following:
>
> 1. Brouwer has, in particular by means of his final circular letter to German mathematicians before Bologna, insulted me and, as I believe, the majority of German mathematicians.
>
> 2. In particular because of his strikingly hostile position vis-à-vis sympathetic foreign mathematicians, he is, in particular in the present time, unsuitable to participate in the editing of the *Mathematische Annalen*.
>
> 3. I would like to keep, in the spirit of the founders of the *Mathematische Annalen*—Klein, who earlier than any of us realized the overall detrimental activity of Brouwer, would also agree with me—Gottingen as the chief base of the *Mathematische Annalen*. (van Dalen 1990, 23)

As can be seen from this quote, Hilbert's reasons were both professional and personal. He accused Brouwer of insulting him and, as he believed, the majority of German mathematicians through a circular letter Brouwer had sent prior to the Bologna conference. Beyond this, Hilbert criticized Brouwer's antagonistic stance toward foreign mathematicians who had supported the journal, claiming this behavior made Brouwer unfit for an editorial role—especially in the turbulent times the academic community was navigating. Viewed from a wider perspective, Hilbert's actions did not occur in isolation; they were shaped by broader political and social contexts, which were in part fueled by Brouwer's own positions and provocations.

7.3.1 The Wider Context: Foundational Debates and Socio-Political Upheaval

During the late 1920s, the academic community in Germany, and Europe more broadly, was grappling with a confluence of political, cultural, and intellectual upheavals that shaped its institutions and the dynamics within them. The aftermath of World War I had left deep economic and social scars, leading to widespread instability in Germany under the Weimar Republic. The rising tide of nationalism and political extremism, particularly the growing influence of far-right ideologies, further polarized German society. These tensions inevitably seeped into academic circles, where scholars were often forced to navigate increasingly fraught ideological landscapes.

In the realm of mathematics, the foundational debates of the time added another layer of contention. The early twentieth century saw mathematics grappling with profound philosophical questions about its nature, structure, and methodology. Competing schools of thought, including Brouwer's intuitionism, Hilbert's formalism, and later, Gödel's incompleteness theorems, sparked intense debates (Hesseling 2003; Mehrtens 1990; Mancosu 1998; Van Heijenoort 1967). These disputes were not merely intellectual; they often became deeply personal, reflecting broader tensions between tradition and innovation, as well as between individual personalities vying for dominance in shaping the field's trajectory.

Compounding these philosophical rifts were the societal pressures on German academia. The country's universities were undergoing significant changes, influenced by economic hardships, political purges, and shifting priorities. The rise of nationalist sentiments began to erode the cosmopolitan and collaborative spirit that had characterized the international scientific community before the war. Foreign scholars, who had previously been welcomed as colleagues, increasingly found themselves treated with suspicion, a trend that Hilbert implicitly referenced in his criticism of Brouwer's perceived hostility toward supportive foreign mathematicians.

The situation at *Mathematische Annalen* reflected this broader turbulence. As one of the most prestigious mathematical journals of the time, its editorial decisions carried significant weight in shaping the discipline. For Hilbert, Brouwer's behavior and philosophical stance represented not just a personal affront but a challenge to the unity and reputation of the mathematical community, which was already under strain from external forces. By advocating for Brouwer's removal, Hilbert sought to preserve what he saw as the journal's role as a bastion of intellectual rigor and collaboration amidst a period of uncertainty and fragmentation.

Einstein's reaction to Hilbert's demand to remove Brouwer from the *Mathematische Annalen* editorial board reveals another complex stance in this highly charged dispute. While Einstein was not particularly sympathetic to Brouwer's temperament or methods, referring to him as a "psychopath" (van Dalen 1990, 24), he was equally reluctant to endorse Hilbert's decisive actions. In his letter to Hilbert, Einstein acknowledged Brouwer's intellectual brilliance, even as he critiqued his erratic and confrontational behavior. Comparing Brouwer's demeanor to that of a court jester granted the freedom to speak and act without repercussion ("Narrenfreiheit"), Einstein advised a degree of tolerance, framing Brouwer as a figure whose eccentricities, though irritating, were perhaps best overlooked.[4]

In a private letter to Max Born, Einstein expressed his view that Brouwer's influence on the *Mathematische Annalen* could have been curtailed through subtler, less extreme measures rather than the outright removal from the editorial board. Yet, despite his suggestions, Einstein made it clear he had no desire to become further involved in what he dismissed as a trivial conflict. He likened the dispute to a "frog-mice battle,[5]" highlighting the absurdity of small disagreements being blown out of proportion. In the context of the Brouwer-Hilbert controversy, Einstein's comparison served to underscore the pettiness of the academic quarrel, suggesting that, like the fictitious battle, the dispute was being granted far more importance than it deserved.

[4] letter from Einstein to Hilbert, 19/10/1928, [JNUL Einstein, 13.141]

[5] Einstein drew here on the imagery of the *Batrachomyomachia*, a satirical epic poem often attributed to Homer. The poem humorously elevates a minor clash between frogs and mice into an exaggerated, mock-heroic battle, borrowing the grand language and style typically reserved for epic conflicts among gods and heroes.

Hilbert's letters justifying Brouwer's dismissal conceal the profound philosophical rift that lay beneath their personal feud. To Hilbert, Brouwer was not just a difficult colleague—he saw Brouwer's intuitionistic approach as a direct threat to the future of mathematics, believing it would lead the discipline to its ruin. The clash, then, was both personal and philosophical, each feeding into the other in a high-stakes academic drama.

The situation escalated when Hilbert failed to secure unanimous support from the other chief editors to remove Brouwer. What began as an internal disagreement spiraled into a legal dispute involving representatives from the Springer publishing house and their lawyers. The outcome was decisive: Brouwer was removed from his position, and the fallout from this episode left deep scars, as from 1928 onward, Brouwer refused to publish in the *Annalen* and dissuaded his students from doing so as well.

The impact of this conflict on Brouwer's work and life was profound, and he never fully recovered the momentum to advocate for his ideas as he had before (Posy 1998). His later publications made only modest contributions to the development of intuitionism, marking a significant shift in his career (Brouwer 1948a, b, 1949). This episode, set against a backdrop of broader political and social upheaval, underscores how the interplay of personal rivalries, philosophical disagreements, and historical context can profoundly shape the trajectory of intellectual movements. At its core, the conflict between Hilbert and Brouwer was more than a professional disagreement; it was deeply rooted in philosophical tensions and clashing visions for the future of mathematics. Yet, it was the personal animosities and the broader political climate that ultimately catalyzed the rupture.

The interpersonal dynamics between Hilbert and Brouwer reveal the fragility of intellectual collaborations when foundational ideologies collide. Hilbert's view of Brouwer as an existential threat to mathematics highlights the psychological dimension of their conflict, one that transformed philosophical debates into personal vendettas. Simultaneously, the political turmoil of the era—marked by instability and a heightened focus on national and institutional identities—played a critical role. The strained social context amplified the stakes of the conflict, intertwining academic decisions with questions of loyalty, reputation, and the preservation of intellectual traditions. Taken together, the personal, political, and social aspects combine to tell a more convoluted story of the evolution of mathematical knowledge, one that comprises the psychological alongside the social.

7.4 Tying Up Loose Ends: Yet Another Philosophy of Mathematical Practice?

To a certain extent, this book fits well within the field of Philosophy of Mathematical Practice (PMP). One of my central arguments is that the evolution of intuitionism cannot be separated from the interpersonal, psychological, and sociopolitical dynamics that motivated the mathematicians who developed it. On the one hand, this situates the thesis within PMP. On the other hand, a closer examination of the field reveals that psychological dimensions of social identity, group belonging, group categorization, and social polarization—as studied in social psychology—are largely absent from PMP. I believe that incorporating these aspects in particular, and the discipline of psychological science more broadly, will lead to a richer and more comprehensive understanding of mathematical practice.

PMP is a branch of philosophy that takes the actual practice of mathematics as its starting point. A broad comparison with mainstream philosophy of mathematics shows that PMP extends beyond traditional concerns with ontology and foundations. It is historically informed, engages with case studies, often incorporates empirical findings, and—most importantly—emphasizes the dynamic nature of mathematics as an activity carried out by human agents (Carter 2024). What makes PMP particularly relevant to this book is its openness to drawing on insights and methods from other disciplines (Mancosu 2008; Carter 2019). In this chapter, that openness is expressed through the integration of social identity theory and group categorization to better understand the development of mathematics and intuitionism in the twentieth century.

Emphasizing the role of human agents in the creation of mathematics, PMP scholars aim to support the view that mathematical knowledge is a form of human knowledge, while still accounting for its distinctive sense of certainty (Hersh 1979; Kitcher 1984; Ernest 1992; Tymoczko 1985). Drawing on case studies of mathematical activity, PMP scholars also consider how mathematics is practiced today—for example, examining what mathematicians themselves say about their motivations for adopting certain principles (Maddy 1992, 1997). In this context, Weyl's intellectual journey offers a rich and layered case study, combining his recorded reflections on the foundations of mathematics with his personal experiences, such as his relocation to Princeton and his growing sense of estrangement from the physics community, a group that had long been central to his professional identity.

A central focus of PMP is understanding what mathematicians actually do when they engage in mathematical work. This emphasis naturally leads to philosophical exploration of mathematical agents, their actions, and the broader nature of mathematical activity itself (Hamami 2023). Under this broad umbrella, philosophers have addressed a wide array of themes, including the structure of mathematical activity (Giaquinto 2005), the role of planning in mathematical reasoning (De Toffoli and Giardino 2014; Larvor 2019; Chemla 2015; Weber and Tanswell 2022), and questions of rationality (Hamami and Morris 2023). Additional areas of interest include the sources of mathematical motivation (Morris 2020), the use of cognitive tools in mathematical work (Ferreirós 2015; Giardino 2018), the character of inferential acts (Larvor 2012), and the way mathematical texts function as action-guiding instructions (Tanswell et al. 2019). Yet even in its more socially oriented strands, drawing on the work of sociologists like David Bloor and Sal Restivo (Bloor 1976; Restivo et al. 1993), this exploration of agency tends to overlook the psychological dimensions of the mathematical agent.

From a psychological perspective, each of the themes that occupy the discussion on agency has a psychological aspect. Rationality in mathematics involves how individual mathematicians make judgments, handle uncertainty, and justify their choices—processes shaped by cognitive biases, heuristics, and personal reasoning styles. Planning in mathematical reasoning involves cognitive processes such as goal setting, anticipation, mental simulation, and problem decomposition, all of which are mental strategies that reflect how individuals organize their thinking over time, manage complexity, and adapt to obstacles, revealing the psychological underpinnings of mathematical practice. Similarly, sources of motivation in mathematics (such as curiosity, aesthetic appreciation, the pursuit of coherence, or the desire for recognition) reflect internal psychological drives that shape how mathematicians choose problems, persevere through challenges, and find meaning in their work.

These topics are among the central pillars in psychological science, each representing a well-established domain of inquiry with extensive theoretical and empirical grounding. Judgment and decision-making research investigates how individuals evaluate information and make choices under conditions of uncertainty, revealing consistent patterns of deviation from logical reasoning due to mental shortcuts and cognitive framing (Tversky and Kahneman 1986; Thaler and Sunstein 2009; Milkman et al. 2009). The related study of cognitive biases and heuristics explores systematic errors in reasoning, showing how people rely on intuitive rules that often lead to predictable mistakes (Kahneman 1982; Pohl 2017). Research on motivation and goal setting focuses on the cognitive and emotional processes that underlie purposeful action, emphasizing how clear, challenging goals enhance performance and how intrinsic motivation supports sustained engagement and creativity (Locke and Latham 2002; Deci and Ryan 2013; Ryan and Deci 2020). Planning and mental simulation studies concern the ways in which people imagine future scenarios, rehearse potential outcomes, and structure their actions to achieve long-term aims—capacities that are fundamental to self-regulation and problem-solving (Taylor et al. 1998; Baumeister et al. 2016; Baumeister and Hutton 1987). The drive for recognition and social validation highlights the importance of social esteem and belonging as motivators of behavior, suggesting that self-worth is closely tied to perceived social value and that people are strongly guided by concerns about how they are seen by others (Leary and Baumeister 2000; Fiske 2018; Sedikides and Alicke 2012).

Despite the relevance of these psychological insights, they have received little to no attention within PMP, which has traditionally emphasized sociological, educational, and cultural dimensions of mathematics (Hersh 2013; Ernest 1998; Lakatos 1976). However, understanding how mathematicians operate in real-world contexts requires looking beyond social structures to include the psychological traits and mechanisms that shape their reasoning and behavior. In Weyl's and Brouwer's stories, this becomes even more evident: Weyl's sense of group belonging affected tremendously his professional identity and eventually shaped the choices he made regarding which scientific directions to pursue (mathematics in favor of physics during his Princeton years). Brouwer's engagement with the significs circles and his recognition of the benefits of group thinking and the asserting character of mathematics for creating collective knowledge have shaped his theory of the creating subject, marking his final contribution to intuitionism. Given PMP's explicit openness to drawing on empirical and theoretical work from neighboring fields (Carter 2019, 2024), findings from psychology should be integrated more fully into its investigation of mathematical activity.

7.4.1 *Looking Ahead: Mathematical Practice in the Age of AI*

The narratives explored in this chapter so far, bring to light a psychological perspective on mathematics that is missing from traditional PMP views. In looking to the future, the rise of new methodologies such as computer-assisted proofs, probabilistic approaches, and artificial intelligence (AI) based tools, reveal another missing aspect from mathematical practice discourse: the integration of AI-based proofs and techniques into PMP frameworks, specifically into discussions about mathematical agency.[6]

Consider for example how the integration of AI into mathematics is changing the status of the mathematician, a figure traditionally seen as the originator and sole author of mathematical knowledge. As AI-based programs generate solutions that are often beyond human anticipation, mathematicians find themselves collaborating with these technologies (Davies et al. 2021; Davis 2021).[7] While these are not necessarily unfamiliar kinds of collaborations, as mathematicians have been known to use computers before (Appel 1984; Parshina 2024), these technologies are now considered to prompt the invention of new methods (e.g. proofs, solutions, examples) and thus of new knowledge (Trinh et al. 2024; Miao and Wang 2024). This shift raises critical questions about authorship, creativity, responsibility, agency, objectivity, and intellectual authority within the mathematical community. If the machine proposes solutions that humans could neither foresee nor produce independently,

---

[6] This section draws on a broader discussion developed in our working paper: "AI-Based Mathematics and Big Science: Characteristics and Implications" (Friedman and Kish Bar-On, in preparation).

[7] There are several perspectives on the nature of collaboration between mathematicians and AI—if we can indeed call it collaboration. At the heart of the debate lies the question of what role the machine plays versus what role the mathematician plays in the process of discovering new proofs. For instance, Davies et al. argue that the work done by the machine and the mathematician are fundamentally different: the machine identifies and analyzes patterns, while the mathematician is responsible for conjecturing and proving. In their study, they hypothesized the existence of a previously unknown relationship between two elements in knot theory. Using a supervised learning model to detect potential patterns, they framed the model as "assisting" human intuition (Davies et al. 2021, 71). Importantly, they designed the model with constraints that prevented it from generating a proof itself, thereby preserving the mathematician's central role. In response, Davis (2021) challenged this interpretation, arguing that the AI tools played only a minor role in the overall analysis—one that could have likely been accomplished using standard statistical methods. He contends that framing AI as "guiding" human intuition is misleading, as the model's main contribution was simply to eliminate many false leads and highlight a few promising ones. The extent to which AI tools influence or shape mathematical intuition is a central theme we explore in more depth in our forthcoming paper (Friedman and Kish Bar-On, in review).

what remains of the mathematician's role as the creative force behind mathematical discovery?

Mathematical creativity is a complex notion, typically described as the capacity to generate original work that meaningfully expands the field of knowledge, introduces new lines of inquiry, and leads to unconventional, innovative, or insightful solutions to existing problems (Liljedahl and Sriraman 2006; Nadjafikhah et al. 2012; Boden 2004; Hadamard 1945). Some of the literature suggests that the foundation of creativity lies in the ability to think about a problem or situation through analogy (Ghiselin 1952). Other studies argue that employing specific cognitive tools such as imagination, empathy, and embodiment can foster creative outcomes (Csikszentmihalyi 1997; Liljedahl and Allan 2013). In all these instances, the underlying theory maintains that the final articulation of a creative idea results from the conscious and intentional efforts of the *human* mind. The integration of AI-based programs and technologies into the mathematical creative process creates a dissonance with respect to ownership and leadership (as well as a conceptual and definition problem): instead of humans leading the creative process, AI-based technologies present solutions that mathematicians did not think of and could not have thought of. To what extent, if at all, can these machine-generated proofs be regarded as creative ideas? If they are deemed creative, then a redefinition of mathematical creativity (and creativity in general) becomes necessary. If they are not considered creative, the mathematical community must determine how to address machine-generated solutions that are undeniably unusual, insightful, and contribute to the expansion of knowledge. Are they deemed non-creative solely because they originate from a non-human agency?

Most discussions on agency in mathematical practice refer to mathematical activity as *human* activity (Pedersen et al. 2005). In these discussions, mathematical agents must possess specific knowledge, skills, and comprehension to carry out the relevant mathematical activities. They are primarily characterized as human agents with typical cognitive capabilities and constraints, such as the ability to use manipulative imagination, execute inferential steps, and comprehend the underlying reasoning behind mathematical processes. While some acknowledge that computers can play a role within mathematical agency, there has been little substantive exploration of how AI systems that generate original and creative proofs (and not only perform calculations) fit into this framework, or the implications of such an integration.

If mathematics is seen as a human creation, deeply rooted in intuition and intellectual effort, the increasing reliance on AI challenges this view. The challenge is not only theoretical or conceptual; it also has methodological consequences. Our fundamental notion of mathematical agency partially shapes how we analyze specific mathematical activities and practices, which, in turn, can generate new empirical data that may refine and reshape our theoretical perspectives. In practice, the mathematician is being replaced by a more collaborative model, where human and machine contributions intertwine. This shift forces us to reconsider the essence of mathematical activity and its connection to human cognition and creativity, challenging the idea that mathematics and mathematical knowledge are rooted or grounded into elementary *human* activities (Kitcher 1984; Ferreirós 2015). At the same time, if we acknowledge machines as active participants in mathematical discovery or even as agents, we must develop a deeper understanding of the nature of the knowledge they generate.

My point here is that discussions about mathematical agency within PMP are currently lacking serious engagement with two topics: (1) the psychological mechanisms involved in mathematical practices, and (2) the integration of AI-based tools into mathematicians' everyday work. While these topics might seem unrelated, there is a close connection between them. Understanding how theories and insights from psychology can be applied to the study of mathematicians' cognitive processes allows for a deeper grasp of mathematical creativity and intuition—both of which are essential for assessing the extent to which AI-based tools function as collaborative components of mathematical practice.

To answer the question I posed at the beginning of this section, this book's approach is not simply another philosophy of mathematical practice; nor is it an argument against PMP. Some readers may consider it part of PMP, and to some extent, that might be the case. However, I am not a big fan of intra- (or inter-) disciplinary boundaries, so I will not be the one to demarcate this book's limits. I believe the more blurred the boundaries between kinds of explanations are, the more they reflect the true essence of our complex social world. This social complexity is evident in every chapter of this book, in every historical episode I have analyzed, and in every mathematician who has been my focus of attention at some point. If there is one take-home message from this book, it is that there is always something more socially complexed than what meets the eye.

References

Abrams, D., M. Wetherell, S. Cochrane, M. A. Hogg, and John C. Turner. 1990. Knowing what to think by knowing who you are: Self-categorization and the nature of norm formation, conformity and group polarization. *British Journal of Social Psychology* 29:97–119.
Appel, Kenneth I. 1984. The use of the Computer in the proof of the four color theorem. *Proceedings of the American Philosophical Society* 128 (1): 35–39.
Atiyah, Michael. 2003. "Hermann Weyl" biographical. *Memoirs* 82:320–335.
Baumeister, Roy F., Kathleen D. Vohs, and Gabriele Oettingen. 2016. Pragmatic prospection: How and why people think about the future. *Review of General Psychology* 20 (1): 3–16.
Baumeister, Roy, and Debra Hutton. 1987. Self-presentation theory: Self-construction and audience pleasing. In *Theories of group behavior*, ed. Brian Mullen and George Goethals, 71–88. New York: Springer.
Bergami, Massimo, and Richard P. Bagozzi. 2000. Self-categorization, affective commitment and group self-esteem as distinct aspects of social identity in the organization. *British Journal of Social Psychology* 39 (4): 555–577.
Berry, J. W. 1980. *Acculturation: Theory, models and some new findings*. Avalon.
Bloor, David. 1976. *Knowledge and social imagery*. Chicago: University of Chicago Press.
Bloor, David. 1991. *Knowledge and social imagery*. 2nd ed. Chicago: Chicago University Press.
Blum, Kevin. 2005. Construction, solipsism, and intuitionistic mathematics. *Macalester Journal of Philosophy* 14 (1):

Boden, Margaret A. 2004. *The creative mind: Myths and mechanisms*. Routledge.
Brading, Katherine. 2002. Which symmetry? Noether, Weyl, and conservation of electric charge. *Studies in History and Philosophy of Science Part B: Studies in History and Philosophy of Modern Physics* 33 (1): 3–22.
Brouwer, Luitzen Egbertus Jan. 1918. Introduction to van Eeden's lecture on intuitive Significs. *BMS* 22B.
Brouwer, Luitzen Egbertus Jan. 1937. Signific dialogues. In *L. E. J. Brouwer - Collected works Vol. 1 - Philosophy and foundations of mathematics*, 447–452. Amsterdam: North-Holland.
Brouwer, Luitzen Egbertus Jan. 1946. Synopsis of the signific movement in the Netherlands. Prospects of the signific movement. In *Collected works: Philosophy and foundations of mathematics*, ed. Arend Heyting, 465–471. Elsevier Science Publishing.
Brouwer, Luitzen Egbertus Jan. 1948a. Consciousness, philosophy, and mathematics. In *Collected works: Philosophy and foundations of mathematics*, ed. Arend Heyting, 480–494. Elsevier.
Brouwer, Luitzen Egbertus Jan. 1948b. Essentially negative properties. *Indagationes Mathematicae* 10:322–323.
Brouwer, Luitzen Egbertus Jan. 1949. *The non-equivalence of the constructive and the negative order relation in the continuum*. North-Holland.
Brouwer, Luitzen Egbertus Jan. 1975. *Collected works: Philosophy and foundations of mathematics*. New York: Elsevier Science Publishing.
Carter, Jessica. 2019. Philosophy of mathematical practice—Motivations, themes and prospects. *Philosophia Mathematica* 27 (1): 1–32.
Carter, Jessica. 2024. *Introducing the philosophy of mathematical practice*. Cambridge University Press.
Chemla, Karine. 2015. Proof, generality and the prescription of mathematical action: A nanohistorical approach to communication. *Centaurus* 57 (4): 278–300.
Cole, Julian. 2009. Creativity, freedom, and authority: A new perspective on the metaphysics of mathematics. *Australasian Journal of Philosophy* 87:589–608.
Cole, Julian. 2013. Towards an institutional account of the objectivity, necessity, and Atemporality of mathematics. *Philosophia Mathematica* 21:9–36.
Cole, Julian. 2015. Social construction, mathematics, and the collective imposition of function onto reality. *Erkenntnis* 80:1101–1124.
Csikszentmihalyi, Mihaly. 1997. *Flow and the psychology of discovery and invention*. Vol. 39, 1–16. New York: HarperPerennial.
Davies, Alex, Petar Veličković, Lars Buesing, Sam Blackwell, Daniel Zheng, Nenad Tomašev, Richard Tanburn, Peter Battaglia, Charles Blundell, and András Juhász. 2021. Advancing mathematics by guiding human intuition with AI. *Nature* 600 (7887): 70–74.
Davis, Ernest. 2021. Deep learning and mathematical intuition: A review of (Davies et al. 2021). *arXiv Preprint arXiv:2112.04324*.
de Haan, Jacob Israël. 1912. Nieuwe Rechtstaalphilosophie. *Rechtsgeleerd Magazijn* 31:480–522.
De Toffoli, Silvia, and Valeria Giardino. 2014. An inquiry into the practice of proving in low-dimensional topology. In *From logic to practice: Italian studies in the philosophy of mathematics*, 315–336. Springer.
Deci, Edward L., and Richard M. Ryan. 2013. *Intrinsic motivation and self-determination in human behavior*. Springer.
Dieterle, Jill. 2010. Social construction in the philosophy of mathematics: A critical evaluation of Julian Cole's theory. *Philosophia Mathematica* 18:311–328.
Dummett, Michael. 1975. The philosophical basis of intuitionistic logic. In *Truth and other enigmas*, 215–247. Cambridge, MA: Harvard University Press.
Eckes, Christophe. 2018. Weyl's philosophy of physics: From Apriorism to holism (1918-1927). *Philosophia scientiae* 22 (2): 163–184.
Ernest, Paul. 1992. The nature of mathematics: Towards a social constructivist account. *Science & Education* 1:89–100.

Ernest, Paul. 1998. *Social constructivism as a philosophy of mathematics*. Albany: State University of New York Press.
Feist, Richard. 2004. Husserl and Weyl: Phenomenology, mathematics and physics. In *Husserl and the sciences*, ed. Richard Feist, 153–172. Ottawa: University of Ottawa Press.
Ferreirós, José. 2015. *Mathematical knowledge and the interplay of practices*. Princeton University Press.
Fiske, Susan T. 2018. *Social beings: Core motives in social psychology*. Wiley.
Fleck, Ludwik. 1981. Genesis and development of a scientific fact. Chicago: Chicago University Press.
Fleck, Ludwik. 1986. The problem of epistemology. In *Cognition and fact. Materials on Ludwik Fleck*, ed. R. S. Cohen and Schnelle Thomas, 79–112. Dordrecht: D. Reidel.
Franchella, Miriam. 1995. Like a bee on a windowpane: Heyting's reflections on solipsism. *Synthese* 105:207–251.
Friedman, Michael. 1999. *Reconsidering logical positivism*. New York: Cambridge University Press.
Friedman, Michael, and Kati Kish Bar-On. A tool or a collaborator? Rethinking mathematical intuition in the age of AI. Under Review.
Ghiselin, Brewster. 1952. *The creative process: A symposium*. Vol. 717. University of California Press.
Giaquinto, Marcus. 2005. Mathematical activity. In *Visualization, explanation and reasoning styles in mathematics*, 75–87. Springer.
Giardino, Valeria. 2018. Tools for thought: The case of mathematics. *Endeavour* 42 (2–3): 172–179.
Gibson, Margaret A. 2001. Immigrant adaptation and patterns of acculturation. *Human Development* 44 (1): 19–23.
Hadamard, Jacques. 1945. *An essay on the psychology of invention in the mathematical field*. Dover.
Hamami, Yacin. 2023. Agency in mathematical practice. In *Handbook of the history and philosophy of mathematical practice*, 1–19. Springer.
Hamami, Yacin, and Rebecca Lea Morris. 2023. Rationality in mathematical proofs. *Australasian Journal of Philosophy* 101 (4): 793–808.
Hersh, Reuben. 1979. Some proposals for reviving the philosophy of mathematics. *Advances in Mathematics* 31 (1): 31–50.
Hersh, Reuben. 1997. *What is mathematics, really?* Oxford: Oxford University Press.
Hersh, Reuben. 2013. *Experiencing mathematics: What do we do, when we do mathematics?* Vol. 83. American Mathematical Society.
Hesseling, Dennis E. 2003. *Gnomes in the fog: The reception of Brouwer's intuitionism in the 1920s*. Basel, Boston, Berlin: Birkhäuser Verlag.
Heyting, Arend. 1978. History of the foundations of mathematics. *Nieuw Archief voor Wiskunde* 26 (3): 1–21.
Hogg, Michael, Dominic Abrams, and Marilynn B. Brewer. 2017. Social identity: The role of self in group processes and intergroup relations. *Group Processes & Intergroup Relations* 20:570–581.
Kahneman, Daniel. 1982. *Judgment under uncertainty: Heuristics and biases*. Cambridge University.
Kirkels, Mireille. 2013. Mannoury's signific philosophy.
Kirkels, Mireille. 2019. *Gerrit Mannoury (1867–1956). Een Relativistisch Denker*. Rotterdam: Erasmus Universiteit.
Kitcher, Philip. 1984. *The nature of mathematical knowledge*. Oxford University Press.
Kreisel, Georg. 1967. Informal rigour and completeness proofs. In *Problems in the philosophy of mathematics*, ed. Imre Lakatos, 138–186. Amsterdam: North-Holland.
Lakatos, Imre. 1976. *Proofs and refutations: The logic of mathematical discovery*. Cambridge, MA: Cambridge University Press.
Lambert, Nathaniel M., Tyler F. Stillman, Joshua A. Hicks, Shanmukh Kamble, Roy F. Baumeister, and Frank D. Fincham. 2013. To belong is to matter: Sense of belonging enhances meaning

in life. *Personality and Social Psychology Bulletin* 39 (11): 1418–1427. https://doi.org/10.1177/0146167213499186.
Larvor, Brendan. 2012. How to think about informal proofs. *Synthese* 187 (2): 715–730.
Larvor, Brendan. 2019. From Euclidean geometry to knots and nets. *Synthese* 196 (7): 2715–2736.
Leary, Mark R., and Roy F. Baumeister. 2000. The nature and function of self-esteem: Sociometer theory. In *Advances in experimental social psychology*, vol. 32, 1–62. Elsevier.
Liljedahl, Peter, and Darien Allan. 2013. Mathematical discovery. In *Encyclopedia of creativity, invention, innovation and entrepreneurship*, ed. Elias G. Carayannis, 1228–1233. New York: Springer. https://doi.org/10.1007/978-1-4614-3858-8_376.
Liljedahl, Peter, and Bharath Sriraman. 2006. Musings on mathematical creativity. *For the Learning of Mathematics* 26 (1): 17–19.
Locke, Edwin A., and Gary P. Latham. 2002. Building a practically useful theory of goal setting and task motivation: A 35-year odyssey. *American Psychologist* 57 (9): 705–717.
Logan, Shay A. 2015. The semantics of social constructivism. *Synthese* 192:2577–2598.
MacKenzie, Donald. 1981. *Statistics in Britain, 1865–1930: The social construction of scientific knowledge*. Edinburgh: Edinburgh University Press.
Maddy, Penelope. 1992. Indispensability and practice. *The Journal of Philosophy* 89 (6): 275–289.
Maddy, Penelope. 1997. *Naturalism in mathematics*. Clarendon Press.
Mancosu, Paolo, ed. 1998. *From Brouwer to Hilbert: The debate on the foundations of mathematics in the 1920s*. Oxford: Oxford University Press.
Mancosu, Paolo, ed. 2008. *The philosophy of mathematical practice*. Oxford: OUP.
Mannoury, Gerrit. 1917. *The social significance of the mathematical form of thought*. Amsterdam, The Netherlands: University of Amsterdam. https://www.math.ru.nl/werkgroepen/gmfw/bronnen/mannoury2.html.
Meertens, P. J. 1956. Sociological conceptions of G. Mannoury. *Synthese* 10:454–460.
Mehrtens, Herbert. 1987. The social system of mathematics and national socialism: A survey. *Sociological Inquiry* 57:159–182. https://doi.org/10.1111/j.1475-682X.1987.tb01040.x.
Mehrtens, Herbert. 1990. *Moderne Sprache, Mathematik: Eine Geschichte Des Streits Um Die Grundlagen Der Disziplin Und Des Subjekts Formaler Systeme*. Frankfurt am Main: Suhrkamp.
Miao, Qinghai, and Fei-Yue Wang. 2024. *Artificial intelligence for science (AI4S)*. Springer.
Milkman, Katherine L., Dolly Chugh, and Max H. Bazerman. 2009. How can decision making be improved? *Perspectives on Psychological Science* 4 (4): 379–383.
Morris, Rebecca Lea. 2020. Motivated proofs: What they are, why they matter and how to write them. *The Review of Symbolic Logic* 13 (1): 23–46.
Nadjafikhah, Mehdi, Narges Yaftian, and Shahrnaz Bakhshalizadeh. 2012. Mathematical creativity: Some definitions and characteristics. *World Conference on Learning, Teaching & Administration - 2011* 31 (January): 285–291. https://doi.org/10.1016/j.sbspro.2011.12.056.
Newman, Max. 1957. Hermann Weyl, 1885-1955. *Biographical Memoirs of Fellows of the Royal Society* 3:305–328.
Niekus, Joop. 2010. Brouwer's incomplete objects. *History and Philosophy of Logic* 31 (1): 31–46.
Parshina, Katia. 2024. Philosophical Assumptions Behind the Rejection of Computer-Based Proofs. *KRITERION–Journal of Philosophy* 37 (2–4): 105–122.
Pedersen, Stig Andur, P. Mancosu, and K. F. Jorgensen. 2005. *Visualization, explanation and reasoning styles in mathematics*. Springer.
Petrilli, Susan. 2009. *Signifying and understanding. Reading the works of Victoria Welby and the Signific movement*. Berlin: De Gruyter Mouton.
Placek, Tomasz. 1999. *Mathematical intuitionism and Intersubjectivity: A critical exposition of arguments for intuitionism*. Dordrecht: Springer.
Pohl, R. F. 2017. *Cognitive illusions: Intriguing phenomena in thinking, judgment, and memory*, 424–445. London: Routledge.
Posy, Carl. 1998. Brouwer versus Hilbert: 1907-1928. *Science in Context* 11 (2): 291–325.
Posy, Carl. 2020. *Mathematical intuitionism*, Elements in the philosophy of mathematics. Cambridge, MA: Cambridge University Press.

Restivo, Sal, Jean Paul Van Bendegem, and Roland Fischer, eds. 1993. *Math worlds: Philosophical and social studies of mathematics and mathematics education*. New York: State University of New York Press.
Roquette, Peter. 2008. Emmy Noether and Hermann Weyl. In *Groups and analysis: The legacy of Hermann Weyl*, London mathematical society lecture note series, ed. Katrin Tent, 285–326. Cambridge, MA: Cambridge University Press.
Ryan, Richard M., and Edward L. Deci. 2020. Intrinsic and extrinsic motivation from a self-determination theory perspective: Definitions, theory, practices, and future directions. *Contemporary Educational Psychology* 61:101860.
Ryckman, Thomas. 2003. The philosophical roots of the gauge principle: Weyl and transcendental phenomenological idealism. In *Symmetries in physics: Philosophical reflections*, ed. K. Brading and E. Castellani, 61–88. Cambridge, MA: Cambridge University Press.
Ryckman, Thomas. 2005. *The reign of relativity: Philosophy in physics 1915-1925*. Oxford: Oxford University Press.
Sady, Wojciech. 2021. Ludwik Fleck. In *The Stanford Encyclopedia of Philosophy, Winter 2021 edition*. https://plato.stanford.edu/archives/win2021/entries/fleck/. Stanford University.
Schmitz, Walter H., ed. 1990. *Essays on Significs: Papers presented on the occasion of the 150th anniversary of the birth of Victoria Lady Welby (1837–1912)*. Amsterdam: John Benjamins.
Sedikides, Constantine, and Mark D. Alicke. 2012. 17 Self-enhancement and self-protection motives. In *The Oxford handbook of human motivation*, vol. 303. Oxford University Press.
Sieroka, Norman. 2007. Weyl's 'Agens theory' of matter and the Zurich Fichte. *Studies in History and Philosophy of Science Part A* 38 (1): 84–107.
Sigurdsson, Skúli. 1991. *Hermann Weyl, mathematics and physics, 1900–1927*. Ph.D. Dissertation, Harvard University, Department of the History of Science.
Sigurdsson, Skúli. 1996. Physics, life, and contingency: Born, Schrödinger, and Weyl in exile. In *Forced migration and scientific change: Emigre German-speaking scientists and scholars after 1933*, ed. M. G. Ash and A. Sollner, 48–70. Cambridge, MA: Cambridge University Press.
Steingart, Alma. 2012. A group theory of group theory: Collaborative mathematics and the 'uninvention' of a 1000-page proof. *Social Studies of Science* 42 (2): 185–213.
Tadmor, Carmit T., Philip E. Tetlock, and Kaiping Peng. 2009. Acculturation strategies and integrative complexity: The cognitive implications of biculturalism. *Journal of Cross-Cultural Psychology* 40 (1): 105–139.
Tajfel, Henri. 1981. *Human groups and social categories: Studies in social psychology*. Cambridge, MA: Cambridge University Press.
Tajfel, Henri, John Turner, W. G. Austin, and S. Worchel. 1979. An integrative theory of intergroup conflict. In *Organizational identity. A reader*, 56–65. Oxford University Press.
Tanswell, Fenner, Brown Joshua, and Bueno Otsávio. 2019. Go forth and multiply: On actions, instructions and imperatives in mathematical proofs. In *Essays on the philosophy of Jody Azzouni*. Cham: Springer.
Taylor, Shelley E., Lien B. Pham, Inna D. Rivkin, and David A. Armor. 1998. Harnessing the imagination: Mental simulation, self-regulation, and coping. *American Psychologist* 53 (4): 429–439.
Thaler, Richard H., and Cass R. Sunstein. 2009. *Nudge: Improving decisions about health, wealth, and happiness*. Penguin.
Trinh, Trieu H., Wu Yuhuai, Quoc V. Le, He He, and Thang Luong. 2024. Solving Olympiad geometry without human demonstrations. *Nature* 625 (7995): 476–482.
Troelstra, Anne Sjerp. 1969. *Principles of intuitionism*. Berlin: Springer.
Troelstra, Anne Sjerp, and Dirk van Dalen. 1988. *Constructivism in mathematics*. Amsterdam: North-Holland.
Turner, John. 1975. Social comparison and social identity: Some prospects for intergroup behaviour. *European Journal of Social Psychology* 5:5–34.

Turner, John C., and Rina S. Onorato. 1999. Social identity, personality, and the self-concept: A self-categorizing perspective. In *The psychology of the social self*, Applied social research, 11–46. Mahwah, NJ: Lawrence Erlbaum Associates Publishers.
Turner, John, and Katherine Reynolds. 2010. The story of social identity. In *Rediscovering social identity: Core sources*, ed. T. Postmes and N. Branscombe, 13–32. Psychology Press.
Tversky, Amos, and Daniel Kahneman. 1986. Rational choice and the framing of decisions. *Journal of Business Ethics* 59:251–278.
Tymoczko, Thomas. 1985. *New directions in the philosophy of mathematics*. Basel, Boston, Berlin: Birkhäuser.
van Atten, Mark. 2004. *On Brouwer*. Belmont: Wadsworth/Thomson Learning.
van Atten, Mark. 2017. The development of intuitionistic logic. In *The Stanford Encyclopedia of Philosophy (Winter 2017 Edition)*. Stanford University.
van Atten, Mark. 2020. Luitzen Egbertus Jan Brouwer. In *The Stanford Encyclopedia of philosophy*. Stanford University. https://plato.stanford.edu/archives/win2017/entries/brouwer/.
van Dalen, Dirk. 1990. The war of the frogs and the mice, or the crisis of the Mathematische Annalen. *The Mathematical Intelligencer* 12 (4): 17–31.
van Dalen, Dirk. 2013. *L.E.J. Brouwer—Topologist, intuitionist, philosopher: How mathematics is rooted in life*. London: Springer.
van Eeden, Frederik. 1908. Is Communisme Een Droombeeld? In *Studies. Vijfde Reeks*, ed. Frederik van Eeden, 222–248. Amsterdam: W. Versluys.
Van Heijenoort, Jean. 1967. *From Frege to Gödel: A source book in mathematical logic, 1879–1931*. Vol. 9. Harvard University Press.
van Stigt, Walter. 1982. L.E.J. Brouwer, the signific interlude. *Studies in Logic and the Foundations of Mathematics* 110:505–512.
van Stigt, Walter. 1990. *Studies in the history and philosophy of mathematics: Brouwer's intuitionism*. Amsterdam: North-Holland.
Weber, Keith, and Fenner Stanley Tanswell. 2022. Instructions and recipes in mathematical proofs. *Educational Studies in Mathematics* 111 (1): 73–87.
Welby, Victoria. 1896. Sense, meaning and interpretation. *Mind* 5 (17): 24–37.
Weyl, Hermann. 1918. *Space-time-matter*. New York: Dover.
Weyl, Hermann. 1922. Die Einzigartigkeit Der Pythagoreischen Maßbestimmung. *Mathematische Zeitschrift* 12 (1): 263–295. https://doi.org/10.1007/BF01482071.
Weyl, Hermann. 1930. Riickblick Auf Zurich Aus Dem Jahre 1930. In *Gesammelte Abhandlungen*, ed. K. Chandrasekharan, vol. I–IV. Berlin: Springer Verlag.
Weyl, Hermann. 1935. Emmy Noether. *Scripta Mathematica* 3:201–220.
Weyl, Hermann. 1939. *The classical groups, their invariants and representations*. Princeton: Princeton University Press.
Weyl, Hermann. 1944. *Meromorphic functions and analytic curves*. Princeton: Princeton University Press.
Weyl, Hermann. 1952. *Symmetry*. Princeton: Princeton University Press.
Weyl, Hermann. 1954. *Algebraic theory of numbers*. Princeton: Princeton University Press.

Index

© The Author(s), under exclusive license to Springer Nature Switzerland AG 2025
K. Kish Bar-On, *Intuitionism Resocialized*, Studies in History and Philosophy of Science 66, https://doi.org/10.1007/978-3-032-05839-3

MIX
Papier aus verantwortungsvollen Quellen
Paper from responsible sources
FSC® C105338

If you have any concerns about our products,
you can contact us on
ProductSafety@springernature.com

In case Publisher is established outside the EU,
the EU authorized representative is:
Springer Nature Customer Service Center GmbH
Europaplatz 3, 69115 Heidelberg, Germany

Printed by Libri Plureos GmbH
in Hamburg, Germany